METAPHYSICAL EXPERIMENTS

posthumanities

CARY WOLFE, SERIES EDITOR

(continued on page 200)

METAPHYSICAL EXPERIMENTS

Physics and the Invention of the Universe

BJØRN EKEBERG

posthumanities 49

University of Minnesota Press
Minneapolis
London

Published by the University of Minnesota Press
111 Third Avenue South, Suite 290
Minneapolis, MN 55401-2520
http://www.upress.umn.edu

Library of Congress Cataloging-in-Publication Data
Names: Ekeberg, Bjørn, author.
Title: Metaphysical experiments : physics and the invention of the universe / Bjørn Ekeberg.
Description: Minneapolis : University of Minnesota Press, [2019] | Series: Posthumanities ; 49 | Based on author's thesis (Ph. D., University of Victoria, 2010). | Includes bibliographical references and index. | Identifiers: LCCN 2018024168 (print) | LCCN 2018056840 (ebook) | ISBN 9781452959511 (e-book) | ISBN 9781517905699 (hc : alk. paper) | ISBN 9781517905705 (pb : alk. paper)
Subjects: LCSH: Science—Philosophy. | Uniformity of nature.
Classification: LCC Q175 (ebook) | LCC Q175 .E3768 2019 (print) | DDC 523.1—dc23
LC record available at https://lccn.loc.gov/2018024168

UMP LSI

No progress can be made in the philosophy of science if the whole settlement is not discussed at once in all its components: ontology, epistemology, ethics, politics, and theology.

—*Bruno Latour*

CONTENTS

ACKNOWLEDGMENTS

Three years of independent immersive research for this project, following a year of intensive comprehensive examination on the canonical texts of Western history of thought, took place under the aegis of the remarkable interdisciplinary PhD program in cultural, social, and political thought at the University of Victoria.

This book would never have come about without the faithful intellectual support of Dr. Arthur Kroker. I am deeply indebted to him for his vast and intrepid scope of interest, his trenchant critical reflections, and for allowing me unparalleled freedom of movement to connect discourses that tend toward isolation.

In developing my thinking about metaphysics, I owe much to my dear friend the philosopher-maverick Seth Asch, who also first led me to read Spinoza. Moreover, the critical perspectives and questions of Dr. R. B. J. Walker and Dr. Stephen Ross sharpened my interdisciplinary thinking, ranging from political science and mathematics to literary theory. Dr. David Cook at the University of Toronto provided important support with his broad understanding of the history and theory of science. Dr. Andrew Feenberg at Simon Fraser University engaged openly with the challenging material from the point of view of philosophy of technology and offered helpful advice in developing the material toward the book that it has now become.

I thank the physicists I met at the European Council for Nuclear Research (CERN) and at Cambridge University, where I first presented my thesis, for their patience with the questions of an outsider. Thanks to Marilouise Kroker and CTheory for kind advice on the publishing process. I am grateful for the critique of peer reviewers at the University of Minnesota Press, as well as the constructive dialogue with editor Cary Wolfe and publisher Douglas Armato, which helped to make this a better book. Thanks also to the diligent work of the editorial team at the Press, including Gabriel Levin and Mike Stoffel.

Funding for this project was provided by a three-year Canada Graduate Scholarship from the Social Sciences and Humanities Research Council.

A NEW WORLD-OBJECT: mirrors being installed on the James Webb Space Telescope. Copyright NASA, Creative Commons.

INTRODUCTION
A Metaphysical Experiment

Metaphysics penetrates all science, for the very simple reason that
it is contained in its point of departure.

—Émile Meyerson

A "world-object," according to the French philosopher Michel Serres,
is a tool commensurable with one of the dimensions of the world—
for example, "a satellite for speed, an atomic bomb for energy, the internet for space, and nuclear waste for time."[1] By this definition, the
new James Webb Space Telescope (JWST), soon to be launched into
space as the most advanced and costly experiment in the history of
cosmology, certainly expands world-dimensions. Designed to measure
the limits of the universe as we know it, the JWST will probe deep into
the history of our cosmic creation since the event known colloquially
as the big bang, the dawn of time according to the prevailing scientific
theory.

Besides the highly specialized technical and astrophysical parameters that define its specific range of operation, this pioneering world-
object is in a deeper sense also designed to interrogate a fundamental
metaphysical question:

What is the universe?

Insofar as this experiment is expected to provide authoritative answers about our cosmos and our own cosmic history, its impending
launch in turn raises the question of how this object comes to know
its own object. What understanding of the universe determines its design, its questions, its parameters? What lies behind the physics of this
world-object?

This book tells the story of how our current idea of the universe came about and how this idea came to determine the scope of mega-scientific experiments such as the JWST. The question of the nature of the universe is metaphysical in the sense that it lies at the very limit of scientific inquiry, where what we really know is not so easily distinguished from what we think we know, believe we know, and would like to believe. Investigating and treading along this treacherous fault line of the knowable, this book is a study of the metaphysical dimensions of historical world-objects in physics that, each in its own way, set key conditions for this latest space telescope. And it is a philosophical critique of the science behind this mission, of our prevailing understanding of the cosmos today.

When the JWST is finally thrust into orbit (at the time of writing, it is scheduled for 2021), after three decades of planning, enormous cost overruns, and a long string of project delays, the hopes and ambitions of cosmologists around the world ride along with it. The giant telescope comes with astronomical risks. When the Hubble Space Telescope was first launched into low orbit around Earth in 1990, it failed to produce workable data, and it took a costly repair mission many extra years to make it operational. But the James Webb will have to operate perfectly on the first try, because its destination—the so-called L2 point, 1.5 million kilometers from Earth, four times farther than the Moon—lies far beyond the reach of rescuing astronauts. The 6.5-meter primary mirror-like surface, nearly three times the diameter of Hubble's, is the largest ever launched into space, and the telescope will rely on many previously untried technologies, such as sensitive light-detecting instrumentation and a new cooling system to keep the spacecraft below the critical threshold of 50 degrees Kelvin—not to mention the risks inherent in the rocket launch itself.

For this reason, one of the most critical constraints in the design process has been to properly test the space telescope for extreme conditions that do not exist on Earth. Because the fully assembled telescope would be far too large to fit into any available thermal vacuum chamber, mission planners had to devise a test protocol that relies on the general method that has guided the development of modern physics for four centuries: break the whole into small parts and model them incrementally, recalculating step by step back toward the whole. According to the JWST project manager: "We aren't only testing—we're also proving our ability to model correctly, which is how we will evaluate the JWST's absolute performance on-orbit."[2] Here, as in all scientific ventures, theory and experiment mutually determine each other.

For the professional astronomers and cosmologists watching from the ground, the risks extend to the future of the discipline itself. Originally budgeted at under $1 billion, the telescope's price tag has continued to grow. Still two years from launch, it is now estimated at $10 billion, meaning that "the JWST has devoured resources meant for other major projects, none of which can begin serious development until the binge is over."[3] By 2010 the telescope's share of the total astrophysics budget in the United States had reached 50 percent, a proportion that was set to grow further amid a general federal funding squeeze. To make matters more volatile, all the space telescopes currently operated by NASA and the European Space Agency, such as the Herschel Space Observatory, which has been positioned at the same L2 point, have reached the end of their lifetimes, thus perhaps generating a feeling of foreboding among some cosmologists that they have placed all their eggs in one basket. If the experiment fails, there will be a dearth of new data to support the future work of astrophysicists and cosmologists.

However, despite the many obvious physical risk factors involved, there is also a less tangible dimension that this book is concerned with, a risk that will persist even if all the technical variables of the megaproject work according to plan. For what is also at stake with the JWST, I would argue, is the risk of paradigm failure, that is, that the current theory of cosmology, which informed every aspect of the experiment design, turns out to have been misguided, if not quite simply wrong. How could that be?

From the outset, an important thing to realize about the James Webb Space Telescope is that it is not really a telescope. In fact, the giant contraptions that today "observe" the outer galaxies do not observe in any optical sense, as they do not photograph the universe directly. Rather, these experimental devices are designed to produce data based on preset detection variables and field delimitations, data that can be reconstructed into advanced imagery. Insofar as the JWST can be called a telescope, it is only as an analogy to the classical instruments that first opened human eyes to the distant universe. The images it creates are fundamentally shaped by preprogrammed search criteria and theoretical parameters, on the basis of complicated predictions derived from the theoretical framework itself. As I will discuss in chapter 1, scientists do not see "through" a telescope, they see "with" it. What does this mean?

To provide a different analogy, cosmologists working with data from the JWST are akin to drivers on an unlit highway at night in a car with broken headlights; all they have to maneuver through the dark, unknown landscape is a GPS map. In lieu of clear vision, they must put their faith in the map provided by a system they themselves designed, a system that is now, in the course of driving, being put to the test. So even if they manage to avoid crashing into a ditch or another vehicle, the question remains, what is it they claim to see, and how do they know it is really there?

The current theoretical framework for cosmology is built on Einstein's general relativity theory. Moderately well demonstrated as valid for conditions within our own galaxy (which in Einstein's time was more or less what was known of the universe), both the theory and its scope have since been transformed exponentially. As I will discuss in chapter 4, Einstein's mathematical framework was married to the hypothesis that the universe developed from a definitive event similar to a nuclear explosion and can be described according to the models of particle physics. The reliance on theories of nuclear physics is significant, because cosmology is based on remarkably few bedrocks of astronomical observation, and the ostensible conceptual unity of the universe has been pushed to extreme scales, with the framework supposed to be valid for galaxies billions of light-years away. The JWST is set to expand the scale even further.

Quantitatively, scale makes little difference, as it in theory can be as infinite as the mathematical limits of the universe itself. But qualitatively, scale matters insofar as vast distances mean that vast implications follow from even the slightest of discrepancies. Tiny fluctuations in observations and measurement of the Hubble constant, for instance, have caused the calculated age of the universe to change by billions of years. Moreover, beyond the accuracy of calculation lies a more problematic claim of the theory: that the universe is composed largely of unknown "dark matter" and determined by "dark energy." Along with several other cosmological concepts, these are theoretical inventions whose principal function is to uphold the coherence of the mathematical framework itself. The crux of today's cosmology is that in order to maintain a mathematically unified theory of the universe, we must accept that 95 percent of our cosmos is furnished by completely unknown elements and forces for which we have no empirical evidence whatsoever. One might think it's possible to question the imperative of mathematical unification, but as I will show in this book, physics is so deeply, historically, and metaphysically dedicated to maintaining theoretical unification that the objective of research is to find evidence for what the framework predicts, in a logic that appears curiously circular.

The astronomer and physicist Michael J. Disney is among a few notable critics from within the field itself who have dared to ask this heretical question: Is the framework of modern cosmology in fact scientific or a glorified kind of mythology of numbers? Disney argues that the current set of theories that constrain and guide the activity of researchers is a shaky edifice based on far fewer actual observations than the number of specially hypothesized parameters used to explain them. This would be an alarming sign for any science, and in this regard cosmology seems to be exceptional. In 2017 a huge controversy in the field erupted when a key aspect of big bang theory, called "inflation theory," was attacked by three senior cosmologists, including one of the original theorists of inflation, questioning its scientific merit along similar lines. This book will attempt to place these debates within a broader context, but from the outset it should be noted that even astrophysicists wonder if they are indeed conducting "science or folktale." Based on his review of the evidence provided by the field thus far, Disney concludes emphatically that "modern cosmology has at best very flimsy observational support."[4] The principal reason for maintaining the big bang theory, he argues, appears to be that it is presently the only alternative.

Because the JWST was designed according to the same theoretical framework, we cannot expect this experiment to change much of the fundamentals. In my driving-at-night analogy, the sudden observation of an unknown passing light is referred back to the GPS screen—that light there must be this thing on the screen here—and if it appears off from the calculations, we can only recalculate the same map. A new sighting inconsistent with the map is not in itself a crisis; on the contrary, for cosmologists it's a problem that becomes an exciting opportunity to recalibrate their variables. But can we question the logic of the map itself without being truly lost in the dark? Without the mathematical framework to guide its operation, in other words, the JWST can make no meaningful "observation."

Thus, the situation for cosmology today resembles the notion from technology studies of "path-dependency," in which current design decisions are so limited by past decisions and old structures that they overdetermine what can actually be made. This is akin to how most institutions over time tend to become more concerned with perpetuating their own existence than serving the purpose for which they were originally set up. Eventually, the cracks in the edifice will give in, threatening the entire cosmological paradigm with failure.

CRITICAL DISCOURSES

Of course, my critique should not be understood as a claim to have a better answer than cosmologists as to what the universe really is. Nor can I offer any alternative scientific cosmology in its stead. The purpose of this book is rather to provide a critical questioning of the limits of our scientific understanding of the cosmos, interrogating fundamentals taken for granted as scientific truths but whose reasons for existence turn out, upon closer inspection, to be more pragmatic, ad hoc, and convenient to other concerns than purely scientific ones. From a critical philosophical viewpoint at the edge of the discourse of cosmology, I try to ask questions that scientists themselves are rarely trained to ask. Moreover, I try to analyze the non-mathematical dimension of their problems as clearly as I can, also for readers whose familiarity with these sciences, or philosophical discourses on their essential problems, may be limited.

Nevertheless, the complexity of the matter is humbling and the perspectives and conflicting passions manifold, as befits a topic that in its appeal to universality has the right to interest so many. Anyone trying to study cosmology, physics, and metaphysics is therefore confronted with a jungle of information, including voices outside the discipline itself, and there is no way to digest it all without ellipses and unintended omissions. My inquiry has been shaped by many influences, notably the philosophers Spinoza, Bergson, Heidegger, and Arendt, as well as the science historians and philosophers Hacking, Gaukroger, Kuhn, and Falkenburg, all discussed throughout the book. However, to situate the critical perspective I bring to bear on this study, I will briefly discuss three significant contemporary strands of thought.

First, Nancy Cartwright's intrepid work *How the Laws of Physics Lie* is a landmark critical perspective on the nature of theoretical physics. Writing as a practicing physicist and philosopher in an anglo-analytical positivist vein, Cartwright critiques the fundamental distinction in physics between phenomenological and theoretical laws. Whereas phenomenological laws are descriptive of concrete and observable situations, theoretical laws are explanatory and function only in the abstract and general (or, in physicists' terms, universal). Compared to other scientific disciplines, physics appears "inside out" in this regard, because it is the only science in which theoretical laws are treated as more fundamental than phenomenological ones. Despite the empirically verified accuracy of observable relationships that phenomenological laws represent, the nature of theoretical laws, bereft of empirical basis, is regarded as the "true," fundamental expression of physics. This is because theoretical laws attempt to explain the deeper reality behind appearances, which is a metaphysical idea par excellence. Cartwright argues for a reversal of this relationship and tries to place physics back on its phenomenological footing, where it has proven itself most successful. However, such a reversal comes at a great cost: abandoning the goal of physicists to discover ultimate theoretical laws.

As I try to show in this book, this quest for ultimate theoretical laws has defined the history of modern physics and cosmology for four centuries. In this regard, Cartwright questions perhaps the most fundamental of metaphysical presuppositions about nature: "Our knowledge of nature, of nature as we best see it, is highly compartmentalized. Why think nature is unified?"[5] For Cartwright this is not a matter of idle speculation, for it concerns the scientific merit of the theories. In her analysis, these theoretical laws are not useful descriptions of reality: "For the fundamental laws of physics do not describe true facts about reality. Rendered as descriptions of facts, they are false; amended to be true, they lose their fundamental, explanatory force." How could this be possible? Providing one example pertinent to this book, Cartwright discusses the law of gravitation, dubbed "the greatest generalization achieved by the human mind." As I will show in chapter 2, Newton's expression of this law (and Einstein's later version of it, discussed in chapter 4) was certainly a major metaphysical achievement. But what about its physical implications?

> Does this law truly describe how bodies behave? Assuredly not. . . . It is not true [as the law says] that for *any* two bodies the force between them is given by the law of gravitation. Some bodies are charged bodies, and the force between them is not [described by the law of gravitation]. . . . This law can explain [gravitation] in only very simple, or ideal, circumstances. It can account for why the force is as it is when just gravity is at work; but it is of no help for cases in which both gravity and electricity matter.[6]

Cartwright's point, which I will elaborate, is that the laws of physics are established on the basis of a logical trick, one in which the actual conditions of the world are first removed before they are recalculated back into the picture. In practice, physics today is composed of myriad overlapping descriptions that clearly work in their pragmatic contexts but do not add up to the unified totality that the theorists of fundamental laws strive for. Therefore, Cartwright concludes, "the laws of physics, to the extent that they are true, do not explain much."[7]

Whereas Cartwright argues from an empiricist perspective about certain branches of physics, my focus is squarely on the physics of cosmology and the metaphysics of its theories and experiments. Thus, this book shares none of the positivist assumptions of Cartwright and contains no mathematical formulas like hers. Instead, I focus on the key assumptions that underlie the mathematical treatment of physical problems.

This is where a second strand of thought has influenced my mode of inquiry more directly. Isabelle Stengers, a philosopher of science with a background in physics, has developed a perspective more closely aligned with mine, informed by Continental thinkers such as Deleuze, Serres, and Latour. In chapter 1, I use Stengers's analysis of the event of the experiment to define, among other terms, the key concept of invention, which describes the reciprocal capture between theory and experiment necessary to make a scientific discovery. In this regard, Stengers's *The Invention of Modern Science* has been more instrumental than the recent *Cosmopolitics,* an essayistic piece that dances around many themes covered in this book, including Galileo, Lagrange, and the rise of thermodynamics.

Stengers's perspective owes much to the influential anthropological work of Bruno Latour in following "science in action," that is, looking at the production of experimental and theoretical statements in the practical context of scientific work. This leads them far from Cartwright's positivism toward a certain idea of constructivism, which should not be conflated with anti-realism or social constructivism:

> The constructivist ambition requires that we accept that none of our knowledge, none of our convictions, none of our truths can succeed in transcending the status of a "construction." It requires that we affirm their historical immanence and that we take an interest in the means invented, and in the authorities invoked, to establish their claim to a stability that would transcend history, taking those means and authorities as constructions that are added to the first. But the constructivist ambition does not require—quite the contrary—that we yield to the monotonous refrain "it is *only* a construction."[8]

Every scientific discovery is in this sense a construction, because scientists inevitably have to mobilize interest, forces, things, and people in new ways to achieve recognition for what can be called a discovery. Stengers employs the term with great respect for what scientists can do and how they can transform the world, and it proves useful in gesturing toward the complex network of variables that go into the work of physics, including the mobilization of capital and politics, which in Latour's view cannot be excluded from an understanding of science. For

Latour, no experiment or theory, regardless of its consistently demonstrated validity, can be considered in isolation from the political, social, and cultural circumstances in which it appeared. As Stengers puts it:

> The actors in the history of the sciences are not humans "in the service of truth," if this truth must be defined by criteria that escape history, but humans "in the service of history," whose problem is to transform history, and to transform it *in such a way that their colleagues, but also those who, after them, will write history, are constrained to speak of their invention as a "discovery" that others could have made.* The truth, then, is what succeeds in making history in accordance with this constraint.[9]

While I share this general perspective on science, my study of cosmology follows a different path than Stengers's and arrives at some different conclusions. For example, whereas Stengers in *Cosmopolitics* treats probability as an ad hoc change in the practical work of physicists, I demonstrate in chapter 3 that the rise of probability could be understood as a new and independent mode of reasoning that caused not only new solutions to old problems but also explains the emergence of some of the new problems that would define physics for the next century. Moreover, while I employ many of Stengers's (and Latour's) key concepts, differences in vernacular are inevitable, as I try to reconcile the language of metaphysics, stretching from Spinoza to Heidegger on the one hand, with the language of physicists from Galileo to Hawking, on the other. Bearing these differences in mind, this book could be read as complementary to the contemporary discourse of science studies.

Third, the critical viewpoint of this project is inspired by a strand of thought called posthumanism. The world-picture of contemporary cosmology exemplifies the historical legacy of humanism, both in orientation and scope. If the idea of humanism at first seems a far cry from a purified mathematical universe, this book will show how the idea of the universe was forged in direct relation to the idea of the human mind as a singular expression of humanity. And in the subsequent development of physics, the stakes of understanding the universe ran in tandem with this notion of the human. The physics of the nineteenth and twentieth centuries claimed not simply to speak for particular physicists from specific cultures, but for humanity as such. To this day, big technological experiments like the JWST or claims to scientific discoveries are framed in terms of the greatness of humanity.

In this sense, my analysis of the emergence of modern cosmology aligns with an insight of Michel Foucault that Cary Wolfe rightly takes as a starting point for the discourse of posthumanism: that the human subject is in fact an invention of history that only emerges in the modern era. Much like the metaphysical inventions I will analyze in this book, the "human" of humanism appears not ahistorically but in response to specific structures and constraints in place and time. "At least since the seventeenth century," writes Wolfe, "what is called humanism has always been obliged to lean on certain conceptions of man borrowed from religion, science, or politics"—and these conceptions, in scientific and religious guises, lie at the heart of the metaphysical inventions this book tries to excavate. As Wolfe points out, the humanism that emerges, cut from Enlightenment cloth, is "its own dogma, replete with its own prejudices and assumptions . . . themselves a form of the 'superstition' from which the Enlightenment sought to break free." Crucially, the notion of the human in humanism is characterized by a flight from the fundamental, affective conditions of human existence itself—it is "achieved by escaping or repressing not just its animal origins in nature, the biological and the evolutionary, but more generally by transcending the bonds of materiality and embodiment altogether."[10]

The history of cosmology belongs to this tradition, as I will show, from Galileo's invention of the means to dispense with the actual world in order to render a universe calculable, to twentieth-century theoretical physics, which consciously attempts to free itself from the shackles of the phenomenal and the observable in favor of a purely mathematical realm of explanation.

As I will argue, the universe is first and foremost a metaphysical idea, an invention. In common parlance, the word "universe" may simply be a name for the vast material reality within which we as earthlings find ourselves, more or less identical to the meaning of "cosmos," the ancient Greek name for a greater order of things. But I will argue there is a significant distinction between these two terms. In the following, I will use cosmos to refer to the general whole within which we as beings find ourselves and universe for the specific scientific idea that emerges in an early-modern historical context.

From its inception, this idea of the universe was shaped by an unstable fault line between believing and knowing, between science and faith, which haunts it to the present day. Despite the dream of empiricists in the early 1900s to demarcate a neat and tidy boundary between the physical and the metaphysical, or between the scientific and the theological, this ideal delineation turned out in the course of the century to be a metaphysical notion in itself. However, at the same time, as I will discuss, the classical philosophical distinction between ontology and epistemology, between being and knowing, was also rendered irrelevant by the developments of twentieth-century physics. For this reason, I use the language of metaphysics rather than ontology, because it provides a more precise analytical meaning in this context.

In other words, not only does metaphysics determine physics, but the work of physics has implications for metaphysics, for how to do philosophy. In my view, discoveries and theories in physics cannot be treated as metaphysical givens and applied back into human sciences without deeper questioning. One of my central objectives with this book is therefore to provide the humanities and social sciences with a renewed insight into the mathematical sciences, which are so influential and yet so opaque.

BRIEF OVERVIEW

I define metaphysics as the limit condition of modern physics, which shows its persistent significance in tacit assumptions, unquestioned principles, hidden implications, overreaching simplifications, and other forms of delimitations that cannot themselves be the subject of a physical experiment but rather figure invisibly as its conditions of possibility. The idea of the scientific universe emerged in the seventeenth century from the premise, much inspired by theology, that the cosmos has a deep unity for which a unified mathematical expression exists and can be discovered. This metaphysical proposition still lies at the heart of the world's costliest science experiments four hundred years later.

In chapter 1, I focus on the metaphysics of a recent mega-experiment, the Large Hadron Collider (LHC) outside Geneva. An underground facility dedicated to exploring the subatomic realm may seem an unlikely starting point for understanding the macrocosmic world, but this experiment has become crucial to cosmologists, because it offers a joint venture with theoretical physicists in pursuit of a "Theory of Everything," a mathematical project with a longer history than its name suggests.

To better understand how nuclear physics and cosmology came to merge their theoretical horizons in the twenty-first century, chapter 2 jumps back in time, to the seventeenth century, to study how the idea of the universe was formed in conjunction with the first telescope. In a close reading of Galileo's role as experimenter and theorist, Descartes's role as metaphysician, and the heretical perspective of lens grinder and philosopher Spinoza, I outline the logical basis for the later Newtonian metaphysics and show how the universe in this sense was an invention.

Following the universal comes the particular, and the scientific invention of the particle in the late nineteenth century would also come to have major cosmological implications. Chapter 3 focuses on the development of physics in a period when it was dominated by phenomena that did not fit the mechanist model and a new probabilistic reasoning whose implications it did not grasp. Looking at the experimental inventions of Maxwell and Planck in particular, along with Bergson's frustrated attempts to reconcile science and metaphysics, I try to show how atomic particles were forged to resolve these emerging schisms. I argue that the metaphysics of the Einsteinian universe—the basic framework for modern cosmology—was born out of an irreconcilable relation between two different modes of reasoning, later resulting in a new divide between relativity and quantum mechanics. To bridge this divide would become the objective of metaphysical experiments such as the LHC and the JWST, which are designed to solicit evidence for a framework in which all physical forces can be combined.

Taking us into the twentieth century, chapter 4 discusses how the big bang paradigm emerged from a new discipline of scientific cosmology in the decades following World War II, when the universe was reinvented on terms that favored giant military-industrial science projects.

I show how Einstein's first four principles of general relativity were morphed by calculations and observations into a new mega-scale theory, consummated by Hawking's idea of a "singularity" as the mathematical proof of big bang metaphysics. From these developments and their implications for how we view the universe today, we arrive back at the coming launch of the James Webb Space Telescope, and in the conclusion I question its relevance as a metaphysical experiment.

Each chapter is framed around a similar historical world-object, an experimental invention crucial to the metaphysics of the time. Each chapter also considers a different dimension of the problems that came to determine the physical understanding of the universe. I define these dimensions as "logics," because I derive them analytically from the core logical principles of Western metaphysics, but my use of this concept should not be understood in terms of an analytic philosophy of logic. Rather, I propose these four logics—analogic, autologic, metalogic, and hypologic—as images of thought that reflect a key metaphysical aspect of the problem to which each experiment speaks. Whether these logics can be turned into a more comprehensive theory is perhaps the subject of another book; in this text, they figure primarily as my own inventions to cast new light on the implications of the metaphysical experiments. And as the metaphorical concept of the big bang already implies, in order to understand what the universe is we must begin with an analogy.

1

FROM MICROCOSM TO MACROCOSM: an experimental particle collision tracker of the Large Hadron Collider. Copyright CERN.

COSMOLOGY IN THE CAVE
The World-Picture of Modern Physics

God is dead: but such as humans are, there may for millennia yet be caves
where they will point to his shadows.

—*Friedrich Nietzsche*

THE ANALOGICAL LEAP

On the outskirts of Geneva, Switzerland, scattered across a campus of
institutional buildings with a large tunnel circling underneath peaceful
farmlands, lies another world-object that succeeded in capturing the
attention of people around the world as it began operations in 2008.
Despite being buried deep underground, it purports to investigate the
key constituents of the cosmos and thereby provide answers to the origin and the end of the universe as we know it. By some measures, the
Large Hadron Collider (LHC) is the most powerful machine ever built,
the largest and (until the JWST is launched) most expensive physics
experiment in history. It is capable of accelerating subatomic particles
up to the speed of light, re-creating theoretical conditions of the big
bang.

In this sense, the LHC is a world-object that also provides what German philosopher Martin Heidegger called a world-picture. Not merely
a picture of the world as such, a world-picture means for Heidegger the
"enframing" of the world, a fundamental ongoing activity that characterizes modern metaphysics in general and physics in particular. "Metaphysics grounds an age, in that through a specific interpretation of
what is and through a specific comprehension of truth, it gives to that
age the basis upon which it is essentially formed," writes Heidegger.[1]
The LHC involves precisely such a grounding for our age, determining

for us the truth of the physical cosmos in which we live. Operating at the limits of the knowable, this enframing is as much metaphysical as physical, because as Heidegger remarks, physics itself is not a possible object of a physical experiment.

Few scientists like to use the word "metaphysics," of course, because they have inherited an attitude from twentieth-century philosophy of science that holds metaphysics as antithetical to properly empirical science. Certainly, practicing physicists on this campus would rather talk about the limited and specific variables of their work. Nevertheless, all ostensibly tangible micro-science conducted at the LHC is affected and delimited by metaphysical assumptions and implications. And what is today called "scientific cosmology" is clearly oriented toward a metaphysical dimension, first because it is drawn from first principles and general assumptions, and second because it operates at the limit conditions of physics as such.

Metaphysics is about limits—the implicit and explicit conditions that enframe the knowable. Limits cannot themselves be subject to testing; rather, they must be implicitly shared (i.e., believed) by all participants. In this sense, all knowing starts with believing, and for practicing physicists, belief pertains to certain fundamental assumptions of scientific inquiry. For Heidegger, metaphysics is primarily formulated through the sciences, because they provide us not only with a definitive (even if continually changing) picture of the world but, more fundamentally, with the projection of the world as a picture in the first place. In this framing activity, metaphysics is not this or that specific picture, but the implicit structuring of the scientific means of articulation. "For the sciences," Heidegger writes, "in manifold ways, always claim to give the fundamental form of knowing and of the knowable in advance, whether deliberately or through the kind of currency and effectiveness that they themselves possess."[2] Herein lies both the most powerful scope of the sciences and its principal blind spot.

As the French philosopher Bruno Latour argues, the sciences today still operate according to Plato's allegory of the cave. In Latour's description, the allegory's "philosopher-scientist," who breaks free from the prisoners' shackles to stare directly at the sun outside the cave, determines the role of the sciences in a decisive double constitution. In the first movement, writes Latour, "the Philosopher, and later the Scientist, have to free themselves of the tyranny of the social dimension, public life, politics, subjective feelings, popular agitation—in short, from the dark Cave—if they want to accede to truth." In the second movement, "the Scientist, once equipped with laws not made by human hands that he has just contemplated . . . can go back into the Cave so as to bring order to it with incontestable findings that will silence the endless chatter of the ignorant mob." In both movements, there is "no possible continuity between the world of human beings and access to truths 'not made by human hands.'" In this double rupture between the cave and its exterior, between fictional shadows and the light of truth, between the social world and the world of nature, the scientist plays a singular role: he "can go back and forth from one world to the other no matter what: the passageway closed to all others is open to him alone."[3] And while the practice of the philosopher-scientist has changed since Plato, the same double rupture prevails. As Latour puts it, "the belief that there are only two positions, realism and idealism, nature and society, is in effect the essential source of the power that is symbolized by the myth of the Cave."[4] And while this philosopher-scientist is a generic figure that can be recognized in many scientific fields, physics has for more than a century laid claim to an exceptional status among the sciences.

As Isabelle Stengers notes, in 1908 the German physicist Max Planck took his positivist colleague Ernst Mach to task for "weakening the faith in the intelligible unity of the world":

> For Mach, physical references that appeared to refer to a world that existed independently—absolute space and time, atoms, and so on—had to be eliminated and replaced by formulations that tied physical laws to the human practices with which they were insolubly connected. In contrast to this critical approach, Planck would affirm the necessity of the "physicist's faith" in the possibility of achieving a unified concept of the world.

At a time when religious dogma no longer posed any kind of threat to the practice of physicists, writes Stengers,

Planck was the first to explicitly position physics within the context of *faith* rather than austere rationality, a faith that had now become an essential component of the physicist's *vocation*, and to correlatively affirm that the practice of physics was not just another kind of science. Planck did not actually deny the general plausibility of Mach's description, he rejected it *for physics*. Physicists *must be able* to speak of the "world" or "nature" independently of the operational and instrumental relationships that, for Mach, were the only source of theory's legitimacy.[5]

More than a century later, physicists and cosmologists still claim to study nature at its purest in ways designed to avoid the vicissitudes of social, cultural, and political conditions. To do their work, scientists must assume that their world-picture constitutes nature in its naked truth. In what does this world-picture most essentially consist?

According to Heidegger, the main current of metaphysics in Western thought since Plato and Aristotle concerns the expression of "what is" according to a basic division of the sensory from the suprasensory. "God" is a name for the "suprasensory world in general," and though the name changes with each historical constellation, the structure is nonetheless maintained:

> If God in the sense of the Christian god has disappeared from his authoritative position in the suprasensory world, then this authoritative place itself is still always preserved, even though as that which has become empty. The now-empty authoritative realm of the suprasensory and the ideal world can still be adhered to. What is more, the empty place demands to be occupied anew and to have the god now vanished from it replaced by something else. New ideals are set up.[6]

Marked by its axiomatic relation between a transcendental truth and nature, the "empty place" is historically carried forth by a series of metaphysical inventions ("new ideals") as the constitutive condition of the scientific universe. In this sense, metaphysics and its ongoing operation through modern scientific activity is deeply related to Platonism in that it still relies on a notion of transcendental truth distinct from the human, sensory, empirical, sentient being. Perhaps nowhere is this bifurcation as deeply ingrained as in modern physics. As I will show, the fundamental metaphysical problems of modern cosmology are in a sense built into the structure of the LHC itself.

Against the perplexity of a commanding world-object making such an authoritative world-picture for our time, we begin with a naive question: How could a local experiment in a tunnel underneath the Alps purport to speak to the nature of the cosmos as such? How could physicists like Planck leap from empirical experiment to the whole universe as such?

In the medieval period, the Scholastics conceived the primary logic by which we can reason our way to the existence of the divine from earthly matters as the *analogical*. The LHC's claims about the cosmos relies on a principal legitimatory link from their local operation through a theoretical framework based on a few key inventions, which I will discuss later. But from the outset, by engaging with physicists' experiments on their terms, we have tacitly accepted the analogical relation between an experiment conducted in a specific place and time and the possibility of a link to the origin of the universe and its end, and consequently everything in between. This analogical link constitutes perhaps the first concealed claim of the experiment itself. Analogy, in its proper etymological sense, is the logic of "ana," the tendency toward inclination. It comes from the Greek word *analogos*, or "proportion"—a relational measure, one thing expressed in terms of another. To set a proportion is to enable a leap between what is being differentiated, relating a part to a whole. Analogy signifies an upward mobility, a leap from one thing in terms of another thing, anything to anywhere. Analogy, in short, reveals the world in its sheer connection. It opens us to the world. Thus, it is where we begin.

Down in a tunnel outside Geneva, analogical relations between space and time, matter and energy, the universe and particles—whose exact configuration govern the world as we know it—constitute and enframe the world according to hidden transcendental ideals. As Heidegger puts it, "within the complex of machinery that is necessary to physics in order to carry out the smashing of the atom lies hidden the whole of physics up to now."[7]

PARADOXICAL SCOPES

Drawing on more than fifty years of research history at CERN (European Council for Nuclear Research), the LHC opened in 2008 and has since been the focal point of advanced physics research across the globe. Composed from the outset of six loosely coordinated experiments, the LHC is as cosmopolitan as the city in which it is situated. As I enter the control room of the ATLAS project, the largest of the collaborations at CERN, I find it reminiscent of a NASA movie set: there are more than a hundred screens, six designated teams working with ten to twenty computer screens each, all facing a long front wall of projected images, data, and command overviews. One team supervises security of the entire system, another controls data input and flow onto the storehouse of servers, and so on, in an integrated model of activities. The room is full of physicists who are fine-tuning and calibrating the complicated system that is set to capture the subatomic particle collisions generated in the tunnel underneath.

ATLAS is centered on a massive, seven-thousand-ton multipurpose detector built around a section of the tunnel. Its contrived acronym (originally short for A Toroidal LHC ApparatuS) invokes the double historical sense of world mapping and, mythologically, bearing the weight of the heavens—inspiration from the activity of giants. ATLAS competes indirectly with CMS, a smaller detector with a similar scope in a different section of the 27-kilometer circular LHC tunnel that cuts underneath the border between Switzerland and France. Both ATLAS and CMS will provide overlapping data, partly for cross-reference and correction and partly to double the extent of possible testing.

Whereas most popular accounts of the LHC focus on its glaring material aspects, such as the sheer enormity of the construction itself, less notice is given to this control room, which is the interface between experiment and physicists. Under what terms and conditions do the thousands of organized researchers come to interpret, analyze, and understand the data displayed before them? Physicists understand the collider essentially as a microscope, but this is also merely an analogy. The ATLAS detector is not a microscope per se but rather is *like* a microscope, because it is built for "seeing into the unknown" of whatever happens in the collider tunnel. In theory, this seems accurate—but whereas the purported objects, the collisions, occur 100 meters below, the observing scientists are deciphering computerized renderings on digital monitors. In what sense does this constitute "seeing into the unknown"?

In the interface between particle collisions and physicists we encounter one of many ways in which twentieth-century physics was forced to pass a critical technological threshold in order to progress. The Canadian historian and philosopher of science Ian Hacking reminds us that from the very beginning of experimental science, observation was associated with the use of instruments. The microscope, in parallel to the telescope, plays a unique role in the history of scientific observation and experimentation for drastically extending the range of scientific inquiry. As a nominal concept for a kind of instrument whose character has changed remarkably over the centuries, the microscope has a history, writes Hacking, marked by three significant, identifiable shifts—or technological jumps. Measured by limits of resolution, we could draw a graph of development that would make its first leap around 1660, then continue along a slowly ascending plateau until a second great leap around 1870; the final major leap, with which the immediate forerunners of the LHC can be directly associated, begins before World War II and continues through the 1950s and 1960s.[8]

These leaps are key markers in the modern history of cosmology, and they will be focal points for the next three chapters. First, the mid-seventeenth-century development of optical instruments, based on the principle of light absorption, coheres with the era that is usually thought of as the birth of modern science, from Galileo's telescope to the Newtonian synthesis of the early eighteenth century. Second, the late-nineteenth-century invention of diffraction microscopes and the optical limit of resolution (perfected by Ernst Abbé and Carl Zeiss) coheres with the growing range of experiments questioning the predictions of the Newtonian paradigm at lower levels of resolution, eventually leading to what we call quantum physics. The first Hubble telescope belonged to this order of instruments. Finally, the mid-twentieth century brought a plethora of highly specialized instrumental machines able to exploit different aspects of light born out of extensive nuclear research, as physics became part of a greater military-industrial expansion. Practically, this meant that microscopes and telescopes could be transformed beyond optical limits, now based on a theoretical framework that quantified resolution in terms of energy. The particle accelerator was first prototyped in the 1930s and built at ever greater scales from the 1950s onward. Insofar as the LHC can be called a microscope, then, it's a third-order invention, much like the JWST, which is a "telescope" by analogy only.

As Hacking puts it, contemporary scientists do not see *through* a microscope—they see *with* it. In the complicated range of images generated by the ATLAS detector, the goal is to observe a track, or a set of tracks, from a collision of particles in the tunnel. One set of tracks is what physicists call an "event." How do they see an event with the microscope? In a sense, they map it. Hacking puts it more generally:

> When an image is a map of interactions between the specimen and the image of radiation, and the map is a good one, then we are seeing with a microscope. What is a good map? After discarding aberrations or artifacts, the map should represent some structure in the specimen in essentially the same two- or three-dimensional set of relationships as are actually present in the specimen.[9]

Note the essential ambiguity of this statement. A "good map" "should represent" what is "actually present," even though the map is all that we know about what is actually present. This bespeaks a problem that defines the modern history and philosophy of science. How does the image, the "good map" generated by the microscope, indicate an underlying reality? In what sense is an event "really" an event?

The traditional answer adopts one of two mutually exclusive positions that are tied to the logic of the cave: the event is either a "human construct" or it is "nature as such." Hacking refers to the two positions as realism and anti-realism; in academic practice they are sometimes referred to as realism and instrumentalism, or realism and constructivism. The discourse is bewilderingly fractured, with a large variance of more ostensibly nuanced positions articulated within this bipolar spectrum—from "structural realism" and "entity realism" through "moderate" and "strict empiricism" to "social constructivism." Moreover, the discourse of scientific realism is historically entangled with an old philosophical fault line between materialism and idealism, which in turn is implicated in the post-Kantian epistemological divide between rationalism and empiricism. And to make matters even more complicated, these debates become inseparable from the early-twentieth-century movement in philosophy that sought to separate the empirical from the metaphysical.[10] Under the guise of logical empiricism, the objective of philosophy became to derive clear rules by which the truth generated by physics could be positively demarcated as scientific. According to this positivist fault line, metaphysics is aligned with theology and must therefore be separated from scientific practice.

Alas, things were never so simple in reality. Beyond the never-ending academic debates, the pragmatics of science add their own twist. As Hacking puts it, "the realism/anti-realism debates at the level of representation are always inconclusive."[11] This is because "whereas the speculator, the calculator and the model-builder can be anti-realist, the experimenter must be a realist."[12] For experimental physicists, the event is a fundamental operational unit, and questioning its reality status amounts to an empty intellectual exercise. Even if the event is not real (a matter of academic debate), it has to be treated as though it is. Thus, the event quickly becomes real through the process, according to a logic I will analyze in more detail in chapters 3 and 4. The experimenter will point to the images on-screen as evidence for actual particles colliding. But what is a particle anyway?

The German philosopher of physics Brigitte Falkenburg demonstrates how the straightforward definition of a particle in Newtonian and Einsteinian physics turns into something much more paradoxical under the prevailing regime of quantum physics. "The 20th century history of the particle concept is a story of disillusion," she writes. "It turned out that in the subatomic domain there are no particles in the classical sense. . . . [A] generalized concept of quantum particles is not tenable either. Particles are experimental phenomena rather than fundamental entities."[13] Through the rise of experimental accelerators such as the LHC, particles have proliferated and become a moving and mutating target. Under the dominant paradigm of physics since the 1970s, called the Standard Model, the number of different particle types has increased to a bewildering constellation, from neutrinos to muons, baryons, and gluons, with the so-called Higgs boson as the final predicted event that the LHC has claimed to find. Since the 1950s, every time a new accelerator operates at new energy levels, new kinds of particle images occur. This is because a particle must exist for an experiment to work, and to show this particle, an experiment is necessary. This fundamental circularity has characterized the development of quantum physics from its inception, and it has been extended into the field of cosmology, which has become increasingly reliant on the theories and methods of particle physics. Falkenburg concludes that neither realism nor anti-realism provides a sufficient view of what takes place in particle physics: "The reality of subatomic particles and quantum processes is not a reality in its own right. Rather, it is relational. It only exists relative to a macroscopic environment and to our experimental devices. The quantum entities [in the microscopic world] are processes, dynamic structures, conserved physical properties, and event probabilities in the macroscopic world."[14]

Relationality implies that particles are constituted by that which defies particularity as such—that which reveals itself to us as "processes, dynamic structures, conserved physical properties, and event probabilities." That is to say, the particles are produced by the experiment. Falkenburg points out how metaphysics structures physicists' activity in decisive ways: for instance, in untestable assumptions about the rational order of phenomena that give rise to methodological principles of unity and simplicity, and in operational idealizations that rely on a notion of nature as independent substance. The paradox is that the more physicists destroy particles, the more they need particle concepts to give meaning to their experiments. The operational definition of the particle, which tacitly stabilizes it as a referential phenomenon, is in this sense conceptual feedback from an experimental process in which particularity is simultaneously presupposed and precluded. (This ability to switch back and forth from practical entanglements to theoretical simplification is analogous to Plato's philosopher-scientist being able to return to the cave with an authoritative "discovery" and bypass the metaphysical complications.)

In turn, this has resulted in severe problems for philosophers trying to understand what is taking place. The inherited fault line from the twentieth century that placed physics on one side and metaphysics on another, Falkenburg argues, resulted in conceptual failure on both sides of the limit. "Epistemologically, empiricism cannot cope with the methods of 20th century physics, whereas ontologically, traditional metaphysics cannot cope with the structure of quantum theory."[15] That is, the development of particle physics and cosmology has gone far beyond the limits of its philosophical defenders and simultaneously undermined the grasp of its possible detractors. Consequently, the grand philosophical dividing line between epistemology (knowing) and ontology (being) is itself put in question. The relation between philosophy and physics today, Falkenburg writes, is therefore best conceptualized as a twofold mismatch. Suspended along the axis of a doubling error, metaphysics occurs neither on one nor the other side of a divide but rather within this divide itself, as an instability field giving rise to a proliferation of paradoxical constructions—such as the ATLAS "microscope" at the LHC.

At the heart of this experiment, the principal role of the microscope-like computer detector system is to digitally rebuild events for analysis after they occur—to provide a detailed image that is a "good map" of interactions involved in the event. The main feature here is selection, what Hacking calls the discarding of "aberrations and artifacts." In the ATLAS detector's "inner tracker," all the charges on the various detector surfaces are gathered and converted into binary signals. But with an estimated one billion collisions per second, the data flow encounters a powerful constraint. This is why the key invention of ATLAS is its data selection system, referred to as different levels of "triggers." The Level 1 trigger, working directly on a subset of information from the other detector components, needs 2 microseconds to make its contingent selection of events—keeping around one hundred of the one billion collisions per second. Thus, 99.9999 percent of the potential events are immediately discarded upon detection. Moreover, the Level 2 trigger further selects and gathers events from the inner tracker, based on Level 1 results, and feeds them into a data acquisition system, where individual events can be reconstructed, according to what interests the researchers involved. An "interesting" event is one that can be used to verify the theoretical object physicists are searching for and is therefore indirectly predetermined by a select set of limitations.

According to ATLAS specifications, the final level of data acquisition stored for subsequent analysis and reconstruction amounts to about one billion events per year. That is derived from an estimated one billion collisions per second. These are rough numbers but still indicative of the procedure: the storehouse of servers at CERN, upon which a highly advanced multi-tier global grid system of distributed computing will be drawing to visualize the microscopic details of each potentially interesting collision, retains only about 1 out of every 32 million events. The event appearing on-screen to the physicist is in this sense already exceptional.

Drawing on ATLAS as our own analogy, we can see the contours of the implicit circularity of modern cosmology and advanced particle physics. Experiments are primarily set up to find evidence for predetermined, hypothesized criteria, and the complicated nature of the experiments, including the vast order of magnitude, means it's not possible to "simply observe" what happens—rather, the results of the experiment are interpreted in light of the theoretical framework within which the experiment is given meaning in the first place. Certainly, the multiple renditions of one experiment can generate consistent patterns and correlations, but this occurs within a predetermined frame that cannot itself be subjected to questioning. The logic may not appear circular to experimenters who believe what they have mapped is real, because they have selected those specimens that best cohere with the theory. But as with any successful invention, it appears in retrospect as self-evident.

METAPHYSICAL INVENTIONS

To an experimental physicist, the event is not so much significant for what it *is* (which is ambiguous) as for what it *does*. In 2012, findings from the LHC were presented to herald the discovery of the so-called Higgs boson, a predicted but previously unverified mechanism at the core of the prevailing regime of physics. The following year, the Nobel Prize for physics was awarded jointly to François Englert and Peter W.

Higgs for the discovery, cementing it as a verified truth in the annals of history. The announcement itself appears as an unequivocal statement of objective fact, but in reality, experimental validation never works independently of theory—and in much of contemporary physics it emerges as a consequence of theory, which allows new realities to be made.

Isabelle Stengers suggests we shift our usual perspective on how modern science operates. In her view, the power of the experiment is not its capacity to speak truth directly—look, there it is!—but indirectly, that is, in its ability to abolish the power of fiction. This is what Stengers calls "the idea of a negative truth: a truth whose primary meaning is to resist the test of controversy, unable to be convinced that it is no more than a fiction among others."[16] In other words, truth is not differentiated according to that claim which is in itself scientific but rather negatively, according to the claims that are deemed nonscientific. This in turn bespeaks a political problem:

> The decision as to "what is scientific" indeed depends on a politics constitutive of the sciences, because what is at stake are the tests that qualify one statement among other statements—a claimant and its rivals. No statement draws its legitimacy from an epistemological right, which would play a role analogous to the divine right of politics. They all belong to the order of the possible, and are only differentiated *a posteriori*.[17]

The experimental statement itself is therefore in principle mute with regard to its positive scope. Only retroactively is it made to speak as nature and truth. In this sense, the experiment plays on a double register of revealing and concealing: it makes the phenomenon "speak" in order to "silence" the rivals. The actual constitution of the event is therefore much more circuitous and counterintuitive than it is later made to appear. In effect, Stengers argues, it's a double constitution: "This is the very meaning of the event that constitutes the experimental invention: the invention of the power to confer on things the power of conferring on the experimenter the power to speak in their name."[18]

In this precise sense, the event is an invention. It introduces a discontinuity whose significance grows through repeated reference to itself—defined, in other words, by its continuing effects rather than its first locus of appearance. In her later work *Cosmopolitics*, Stengers refers to this double constitution as "reciprocal capture." As soon as the collider and tracker apparatus begins producing its highly select data, scientists involved work to further select their collision-event, by differentiating it within a set of theoretical predictions with which they have already engaged. Could this particular event be the one they search for, such as the Higgs? They rebuild the event digitally, they analyze it. They produce publications through reference to it, which become echoed in more publications. They may encounter rival claims and rival interpretations, based on other events analyzed along divergent lines, but this only means they have succeeded in garnering interest for their event. The more controversy, the more interest. As more physicists are engaged and tacitly confer power on the experiment and its discovery, a complex reverberation process develops, which extends to news stories, popular science articles, blogs, generating public interest and assigning roles, according to the ongoing double demarcation of fiction from non-fiction and science from non-science. Through its effects, the event becomes retroactively constituted as exceptional and conceived as an origin, a foundational point in space-time—a difference that makes a difference through a process of differentiation. And one day, the scientists can come back into the cave and announce: "We found it!"

A key concept for the successful creation of an event is therefore interest. As Stengers points out, there is an essential ambiguity to the notion of interest that bespeaks the paradoxical practice of the modern sciences. On the one hand, the denunciation of interest defined in political economic terms is what defines scientists. Their "disinterest" as regards "outer" influence is critical to their status—in direct analogy to the external limit that marks non-science from science. They are objective, disinterested. On the other hand, the movement of scientific invention, the creation of events that make history, depends precisely on interest in a more direct sense. For it is imperative for physicists to succeed in generating interest for their event as differentiated from others.

Etymologically, as Stengers reminds us, to have or enact interest is to be *inter-esse*—situated in between. There are two different ways of understanding what it means to be "in between." On the one hand, interest can be defined as a position in an "external" political-economic situation, which scientists must strenuously seek to avoid. In this sense, to be in between means being an intermediary, caught in between existing forces or things. Scientists are thus conceived as independent points, or a cluster of points, whose exceptional status relies on being able to step outside the field of forces within which they act. On the other hand, interest can be defined as making a link between forces or things in the "internal" sociopolitical situation that acting scientists necessarily are in. Here, to be in between means being a mediator that opens up new relations.[19] Metaphysically, a mediator differs from an intermediary in that it cannot exist independently. Whereas an intermediary is in a symmetrical relation to its surrounding forces, equidistant from either side, the mediator maintains an asymmetrical relation to its constitutive forces: it is itself constitutive of its external relation. Interest in the sense of mediation is therefore not really a position as much as an action that changes the relations through which it becomes expressed. The defining feature of mediation is its double power, or its double movement of power, which not only creates the possibility of translation but also "that which" is translated, insofar as it is capable of being translated and making history in a new sense.

Mediation is complex, and it is no wonder if scientists (and the public) prefer the idealized image of the intermediary that flattens the process and preserves the shine of authority. The much lauded Higgs event, for example, makes a certain kind of history, in which the physicists involved appear properly disinterested and objective. But simultaneously, the event signifies for them a chance to conduct further research and take part in the prestige and glory of "Nobel Prize" science, and so on. To the enterprise of particle physics, a successful discovery is of paramount interest, as it both confirms a research program based on several decades of investment and legitimates a claim to conduct new research. The prospect of an end to further research is a true scientific crisis and therefore rarely happens in practice. Instead, the resolution of a problem always leads to another problem.

In the case of the Higgs boson, its experimental discovery appears to verify the entire comprehensive framework for particle physics, the Standard Model, which is built around perhaps the most foundational of problems in modern physics—the paradox of force, or the ostensible impossibility of integrating all known physical forces within the same theory. The Standard Model effectively bifurcates the entire material content of the universe into two different kinds of subatomic particles: "matter" and "interactions," or in the vernacular of physics, fermions and bosons. Physicists differentiate them according to their respective statistical frameworks: fermions follow Fermi-Dirac statistics, and bosons follow Bose-Einstein statistics. As I will discuss in chapter 3, this order of physics implies that all particles are primarily statistical phenomena. Metaphysically, bosons and fermions are distinguished according to an exclusion principle of quantum physics: several bosons, or interaction particles, can occupy the same quantum state, but only one fermion, or matter particle, can occupy the same state, or have the same energy, at any given time. This model is designed to unify three out of four known separate forces in the universe: the gravitational, electromagnetic, weak, and strong forces. The strong force is a nuclear force, explicated by the Standard Model in terms of its own set of particles. After the LHC appears to have yielded results in support of the projected Higgs boson, it marks the theoretical consolidation of the strong force with the previously combined electromagnetic and weak forces. This makes the Standard Model a so-called Grand Unified Theory of physics, or a GUT.

Nevertheless, the GUT that now appears vindicated still leaves one major force beyond the scope of the theory—and this becomes the projected future path of particle physics research, which aligns its interests with cosmology. For the Standard Model fails to account for gravity. Technically, gravity appears in an order of magnitude far below the other forces and only becomes significant on cosmic scales, where in turn it becomes all the more crucial—indeed, a return to the central problem of modern physics.

In ancient, medieval, and early modern thought until Newton, gravity was conceived as a force that appeared as a problematic given, a certain definitive presence that nevertheless appeared incalculable.

Historically, this kind of force was conceptualized as substance—that which is cause of itself, existing in itself and through itself. With thermodynamics and relativity theory, force became energy, and energy is, as the German physicist Werner Heisenberg once pointed out, only a different name for the same thing:

> Energy is in fact the substance from which all elementary particles, all atoms and therefore all things are made, and energy is that which moves. Energy is a substance, since its total amount does not change, and the elementary particles can actually be made from this substance as is seen in many experiments on the creation of elementary particles. Energy can be changed into motion, into heat, into light and into tension. Energy may be called the fundamental cause for all change in the world.[20]

The conversion of force into energy and its subsequent means of quantification make the unification of forces appear as a necessity. We can convert their levels of energy and place them on the same scale, so why can we not create a theory that can account for them all on the same terms? In fact, the possibility of quantifying force belies significant qualitative differences. Electromagnetism, for example, exhibits both an attractive and a repulsive quality, whereas gravity, according to general relativity theory, is purely attractive and hence accumulative, increasing with greater mass in a nonlinear relation that inevitably tends toward collapse.[21] For centuries, the conception and calculation of gravity have caused immense problems for physicists, as I will discuss further in chapter 4, and more advanced means of quantification have not changed the fundamental nature of the problem.

However, few would take seriously the suggestion that mathematically incompatible physical forces in our cosmos might mean they are different natures altogether. Instead, the quest is on, and has been going on for decades, engaging thousands of physicists in work on theories that can unite the Standard Model with gravity and merge the quantum physics of atomic structures with the order of stars and galaxies. This movement would constitute the leap from a GUT to a TOE—a Theory of Everything, the holy grail of theoretical physics, in which all known physical forces can be accounted for within the same framework.

The driving force behind theoretical unification is deeply embedded in the enterprise of physics itself, its logic and its history. A Theory of Everything is in this sense the current operational term for what in the seventeenth century was termed *mathesis universalis*. Its classical definition was offered by the French philosopher René Descartes:

> There must be a general science that explains everything that can be raised concerning order and measure irrespective of the subject matter, and . . . this science should be termed *mathesis universalis*—a venerable term with a well-established meaning—for it covers everything that entitles these other sciences to be called branches of mathematics. How superior it is to these subordinate sciences both in usefulness and simplicity is clear from the fact that it covers all they deal with.[22]

At stake in the reunification of physical forces is therefore the act of making all physical phenomena conform to ultimate laws authoritatively described by the sovereign branch of pure mathematics. This speaks to a peculiarity of physics as a modern science that cosmology has inherited. As Stengers puts it:

> On the one hand, this is clearly the science where the relation between theory and experience is the most rigorous and demanding. . . . But, on the other hand, this is a science that always appears to involve the project of judging phenomena, of submitting them to a rational ideal. More precisely, we are dealing with the only science that makes the distinction between what physicists call "phenomenological laws" and "fundamental laws." The first may well describe phenomena mathematically in a rigorous and relevant way, but only the second can claim to unify the diversity of phenomena, to go "beyond appearances."[23]

This echoes Nancy Cartwright's incisive critique, discussed in the introduction, of the oddity that distinguishes physics from other sciences. In the case of gravity and the Standard Model, we now have a set of "phenomenological laws" that can explain all known physical phenomena—and yet the only accepted form of progression for physics is to go "beyond appearances" and find a mathematical model that can be accepted as a fundamental law. Somehow, the world-picture of physics is turned inside out, and in the next chapters I will try to chart some of the logical flips in its history that have come to determine this curious situation.

Nevertheless, the chances of generating definitive events with the LHC that could lend evidential support to conjectures beyond the Standard Model are slim. By one calculation, a machine powerful enough to generate data for some predicted constituents of a truly unified theory would require fifteen times the electron-volt magnitude of the LHC; by another, several thousand times the energy (and research funding) will be needed. The unlikelihood of testing the theory in a machine like the LHC has led physicists toward a different horizon of possibility—and this is where physics and cosmology become conjoined.

UNIVERSAL THEORIES OF EVERYTHING

From a purely mathematical perspective, the unification of all forces into a TOE is already accomplished. It is called string theory (or in some versions, brane theory), and it derives from an attempt to shift the definition of the smallest constituent of nature from the now destabilized particle concept into something more pliable, an extendable range of tension.

Quantitatively, the finitude of the string is provided by what is called the Planck length, the lowest definable limit in quantum physics, the so-called Planck constant: 10^{-35} (0.0000000000000000000000000000000000001) meter. The metric finitude of the string is not so much an expression of an actual physical size as a lower conceptual threshold toward which string theory must be mathematically coherent in order to properly constitute a Theory of Everything. The idea of string theory, in other words, is to derive a mathematical expression for all physical conditions from its own ultimate finitude.

Significantly, the movement toward string theory does not merely characterize a transgression of the classical physical theories—it also constitutes an inversion of how physicists ask questions about nature. Well into the last century, the physicist used microscopes and telescopes in conjunction with mathematics to ask what fundamental physical entities actually exist within a given universe. Around the turn of the twentieth century, for example, the dominant problem of physics concerned whether the ultimate constituent of nature was an atom or an omnipervasive field of ether. With the later emergence of particle accelerators and high-energy microscopy, new kinds of particles derived through experimental invention proliferated and effectively undermined any semblance of metaphysical unity. Today, the theoretical physicist instead asks: What are the conditions under which mathematical universality is still possible?

The original string theory, developed in the 1970s, had an elegant answer that could fit all known physical phenomena within its limits—as long as the universe consists of twenty-six dimensions. In the later M-theory, which links supersymmetry with quantum field theory, these twenty-six dimensions were progressively compacted into eleven, defined as ten spatial and one temporal. In other words, there is a unified theory that can account for all known physical forces and constitute a fundamental law of nature, but only on condition that the world as we know it contains seven entirely unknown dimensions for which we have no empirical evidence whatsoever. The progression of scientific inquiry is to turn this ostensible absurdity, the obvious discrepancy between theory and reality, into an object of further research. The universal symmetrical perfection of string theory is broken up under local conditions of symmetry breaking, in order to reduce eleven dimensions to the four we know, and then to give all the supersymmetric postulated particles mass and differentiate all physical interactions, so that we can explain how the extra seven dimensions that model postulates are "in fact" hidden from us. The four dimensions we know are "phenomenological," but in a fundamental law "beyond appearances" the universe must consist of eleven dimensions.[24] The world must conform to the model, predicated on an unquestionable mathematical unity. Thus, some of the world's most brilliant minds are dedicated to solving this vexing problem of how the eleven-dimensional world hides its true symmetrical-theoretical perfection inside itself and only appears to us as our imperfect material four-dimensional world. Mathematically, in other words, the problem was solved a long time ago.

However, the true challenge lies in transforming reality to fit the math. When theoretical physicists proposes a model of string theory, it effectively becomes an event that follows a dynamic similar to that of the experimental event. Insofar as theorists can garner interest among peers for their formulations, they contribute to a complex collective movement that ensures, if successful, a shifted practical course for newly interested physicists. Internally, in physics circles, string theory has caused controversy among those who remain faithful to the twentieth-century positivist distinction and thus denounce the theory as "metaphysical" as long as it cannot be tested, in turn yielding creative suggestions for how to shift the criteria for scientific validation, and so on. Nevertheless, the metaphysical thrust of string theory lies not so much in its unverifiable predictions as in the implicit way it structures scientific inquiry. The claim may appear to be that the world actually consists of eleven dimensions, and the controversy around this statement is part of the event's "effects," its unfolding; it succeeds in attracting interest and reinforces a much more circuitous, precedent claim that mobilizes physicists to transform the world in accordance with the demands of the theory. In this sense, the task of theoretical physicists is to mobilize their experimentalist peers into designing tests that could indicate the existence of these unknown dimensions and symmetries. New kinds of tests are theorized, and criteria and parameters are invented. With metaphysical inventions like string theory, a new experimental reality must be produced in order to become retroactively "discovered" as a ground for new theories and experiments. More caves, more shadows, more suprasensory ideals to be substituted for one another—and of course more research.

At the LHC, the ATLAS team is hoping to generate some plausible signs of supersymmetry, a theoretical model that is a necessary condition for string theory. Supersymmetry predicts a hidden coupling between bosons and fermions, in effect positing a universal doubling, in which every known boson and fermion has a hitherto unknown superpartner. If the ATLAS team can find events among the trillions of particle collisions in the machine that do not conform to the Standard Model, it may be read as a sign of such hidden particles, which in turn may allow the mutually exclusive relationship between bosons and fermions to be reordered under a new and expanded regime. And so the search goes on.

One of the analogical claims of the LHC that interests cosmologists is that it can reproduce conditions in the early universe, according to the big bang theory of nucleosynthesis, which gives cosmologists a chance to search for further clues to their theoretical framework. For theoretical physicists, however, the logic works the other way around. If the particle accelerator is a big bang machine, then insofar as prevailing cosmological theories are correct, the universe itself is the ultimate particle accelerator, able to provide the energies, temperatures, and densities high enough to probe the experimental depths of string theory. Without capital for a new machine on Earth, a proper alignment of theoretical physics and cosmology could mean that observation of exceedingly distant galaxies could be taken as evidence for a theory of the ultimate nature of things. In 1993, results from an experiment at CERN in Geneva appeared to confirm predictions about the neutrino particle that were derived from a cosmological argument. An American physicist involved in the experiment, David Schramm, argued that "this was the first time that a particle collider had been able to test a cosmological argument, and it also showed that the marriage between particle physics and cosmology had indeed been consummated."[25]

Could "seeing into the universe" with a telescopic apparatus simultaneously constitute seeing events of a microscopic order? As astronomer and science writer David Lindley puts it: "The hopes of cosmologists and particle physicists have become the same method: on both sides of this joint effort there is absolute reliance on the notion that a single theory will explain everything, and that when such a theory comes along it will be instantly recognizable by all. This might be called the messianic movement in fundamental science."[26] In forging the analogy between a local laboratory and a universal reach, a future beyond the LHC involves the ultimate metaphysical inversion: turning the universe itself into a physics experiment, searching galaxies billions of light-years away for events that fit with the current *mathesis universalis*.

String theory, as only the latest in an array of historical world-pictures, thus fits neatly into Heidegger's musings on Western metaphysics. Heidegger attempts to articulate how the limit condition of metaphysics constitutes a planetary danger claiming the living world.

He has in mind a "metaphysical danger" for which the very emergence of Big Science in his mind is symptomatic. "What claim do we have in mind? Our whole human existence everywhere sees itself challenged—now playfully and now urgently, now breathlessly and now ponderously—to devote itself to the planning and calculating of everything."[27] To this historical challenge Heidegger addresses his metaphysical questions about "the Atomic age" and in particular his insight on the relation between the metaphysical and the mathematical. In *What Is a Thing?*, a work on the metaphysics of modern science with particular emphasis on how the view of the physical world changed from Aristotle to Newton, Heidegger writes that the mathematical is in etymology and essence a mode of learning—that which is learnable. *Mathesis universalis* in its original sense means the learning attitude by which things are taken up in modern knowledge. The common interpretation of mathematics as dealing with number is according to Heidegger only an expression of the mathematical in this deeper sense:

> The mathematical is that evident aspect of things within which we are always already moving and according to which we experience them as things at all, and as such things. The mathematical is this fundamental position we take toward things by which we take up things as already given to us, and as they must be and should be given. The mathematical is thus the fundamental presupposition of the knowledge of all things.[28]

If the mathematical stance implies a configuration of the world such as to reveal itself in terms of learnable things, it is what rules and determines the basic movement of science itself.

Grappling with the fundamentally different characteristics of classical Newtonian physics on the one hand and atomic physics on the other, Heidegger contextualizes his thesis. These two forms of physics, he observes, indicate an epochal shift within modern physics itself, wherein it constitutes and determines nature in two incommensurable ways. And yet, as he points out, "what does not change with this change from geometrizing-classical physics to nuclear and field physics" is the way nature has to be already set in place as knowable "object-ness," or enframing. Whether physics is understood in terms of geometry or statistics, nature is always encountered as an object, even if, as in the case of quantum physics, it is retroactively constituted in its trace. In Heidegger's overview of the physics of his day, he adds a rather curious remark:

> However, the way in which in the most recent phase of atomic physics even the object vanishes also, and the way in which, above all, the subject-object relation as pure relation thus takes precedence over the object and the subject . . . cannot be more precisely discussed in this place.[29]

The vanishing object and the pure relation—this is as good a philosophical description of the mathematical string as any. In cryptic brackets, Heidegger intimates that this pure "relational" ordering of subject and object, thinker and thing, is now itself taken up as standing-reserve— that is, it becomes a new part of the way the world is enframed for scientific inquiry.

Once the paragon of probing the things themselves, today the science of physics rather fits a Platonic-Christian image of a twenty-first-century priesthood of physicists, proselytizing a higher-dimensional world whose access is explicitly contingent upon accepting their metaphysical configuration of *mathesis universalis*. The enduring power of metaphysics lies not simply in these transcendental claims themselves, which can be individually debated and refuted term by term, but rather in how the mode of scientific inquiry is configured by a certain set of hidden assumptions. How could a once "solid" science of physics turn into the most "metaphysical" of all scientific disciplines? How did this profound reversal of asking questions about nature occur?

In the following I will show how the current paradigm of cosmology emerges out of a longer metaphysical tradition. As a brief history of microscopes and telescopes suggests, this development has occurred in leaps and involves some crucial inventions that make it possible to speak of the contemporary situation as an evolution of a linear history. And yet in other important respects, today's cosmological physics emerges as something altogether different, a radical break with its own past. In the early twenty-first century, physics has reached the limit of its "third-order" solution and is once again forced to surpass it if it is to both perpetuate itself and resolve its fundamental problem. But as the next chapters will suggest, perhaps the deeper problems in the universe of modern physics are found in the history of its own making.

2

MODERN WORLD-OBJECT: Galileo's public demonstration of his new telescope, 1608. Copyright Wellcome images, Creative Commons.

OF GOD AND NATURE
The Invention of the Universe

I should wish to demonstrate by certain reasoning things that are contrary to reason.

—*Benedict Spinoza*

THE AUTOLOGICAL MEDIUM

In 1609, Galileo Galilei peered through his new invention for the first time. Drawing on existing optical designs that could magnify objects up to three times their size, Galileo was able to increase this factor to twenty. Suddenly, an entire new dimension of the world was in sight. As a metaphysical invention, his telescope was not simply about enlarging objects in the sky; more significantly, he forged a perspectival shift that could consider the Earth from the viewpoint of the universe. As the philosopher Hannah Arendt argues:

> What Galileo did and what nobody had done before was to use the telescope in such a way that the secrets of the universe were delivered to human cognition "with the certainty of sense-perception"; that is, he put within the grasp of an earth-bound creature and its body-bound senses what had seemed forever beyond his reach, at best open to the uncertainties of speculation and imagination.

This perspectival shift became pivotal to the later Newtonian universe. And as Arendt points out, writing well before the age of satellite Earth mapping, Galileo's telescope functions as an "Archimedean point" through which it becomes possible to act within terrestrial nature as though we are disposing of it from the outside.

Whatever we do today in physics—whether we release energy processes that ordinarily go on only in the sun, or attempt to initiate in a test tube the processes of cosmic evolution, or penetrate with the help of telescopes the cosmic space to a limit of two and even six billion light years, or build machines for the production and control of energies unknown in the household of earthly nature, or attain speeds in atomic accelerators which approach the speed of light, or produce elements not to be found in nature, or disperse radioactive particles, created by us through the use of cosmic radiation, on the earth—we always handle nature from a point in the universe outside the earth.[1]

In this sense, Galileo's telescope was for Arendt also a pivotal moment in the irreversible and paradoxical modern process of "world alienation": the more humans increase their surveying capacity, the more the actual place from which this surveying takes place disappears to them. "Any decrease of terrestrial distance can be won only at the price of putting a decisive distance between man and earth, of alienating man from his immediate earthly surroundings."[2]

This new world-object would cause several unpredictable effects. For the history of philosophy, one of the most significant consequences turned out to be a book that never appeared.

In 1633, a promising young French natural philosopher named René Descartes was on his way to his regular bookseller in Amsterdam. Living in the Netherlands, the most tolerant state at the time for freethinkers, Descartes had recently finished writing what he considered a revolutionary treatise called *The World*, which presented a bold hypothesis on the nature of light, motion, and the dynamics of the universe, clearly inspired by the new vistas of the universe that inventions such as Galileo's had brought about. As Descartes had written to his longtime correspondent Marin Mersenne just a few years earlier:

> instead of explaining a single phenomenon, I have decided to explain all natural phenomena, that is, the whole of physics. And the plan gives me more satisfaction than anything previously, for I think I have found a way of presenting my thoughts so that they satisfy everyone, and others will not be able to deny them.[3]

But when he got to the bookseller, Descartes's aspiration to compose such an "undeniable" natural philosophy took a sudden and unexpected turn with some upsetting news. Anguished, he hurried back home and wrote back to Mersenne:

> I had intended to send you *The World* as a New Year gift . . . but in the meantime I tried to find out in Leiden and Amsterdam whether Galileo's *World System* was available. . . . I was told that it had indeed been published, but that all copies had been burned at Rome, and that Galileo had been convicted and fined. I was so surprised by this that I nearly decided to burn all my papers, or at least let no one see them. . . . I must admit that if this view [that the Earth moves] is false, then so too are the foundations of my philosophy, for it can be demonstrated from them quite clearly. And it is such an integral part of my treatise that I couldn't remove it without making the whole work defective. But for all that, I wouldn't want to publish a discourse which had a single word that the Church disapproved of; so I prefer to suppress it rather than publish it in a mutilated form.[4]

Descartes's crossing of paths with Galileo, now an official heretic of the Catholic Church, had dramatic consequences for himself and the history of metaphysics. As his distraught quote intimates, Descartes was plunged into a crisis of faith. He held off on publishing his grand work, recoiling instead into a renewed and intensified search for methodological justification. How could he properly ground his new vision of the world without being accused of heresy?

Meanwhile, outside Florence, a convicted Galileo under house arrest was also spurred toward methodological justification. Here, he secretly wrote his final work, the *Discourses concerning Two New Sciences*, to be smuggled out of Italy and published in the Netherlands in 1638.[5] The book laid out the novel kinematic rationale that would become the explanatory basis for Newton's dynamics and in turn ensure, in no small way thanks to his legendary public clash with Pope Urban VIII, Galileo's place in history as the father of modern physics. The revolutionary new cosmology of heliocentrism was already developed, in large part by Copernicus and Kepler, but the theory was still lacking affirmation within the existing Christian worldview. What was needed was a decisive grounding for the new way of seeing the world. For Descartes, a Catholic still living in Protestant northern Europe, the intense thinking process would culminate in the theory of the *cogito*, the mental subject at the basis of what would become modern epistemology, in turn ensuring Descartes's role as the father of modern philosophy. In the oblique encounter between these two characters and their two books—one censored by the church, another self-censored by personal faith—lies a mutual origin story for what we now call modern science and philosophy.

Seventeenth-century Europe constituted on almost all accounts a remarkable historical contraction that gave rise to a proliferation of new phenomena. The Thirty Years' War (1618–48), a war of imperial power and religion fought partly by navies and in colonies, with eight million casualties from at least three different continents, can make a plausible claim to being the first real world war. This violent age saw the emergence of, among other things, revolutionary military mobility in the form of musketeer artillery forces; the first stock market; a commercial banking system; currency inflation; an international legal system for the absolute sovereignty of the nation-state; and not least the invention of probability reasoning and statistics, which would later incur a metaphysical revolution on its own. In other words, in the mid-seventeenth-century we find a clustering of war, capital, and science in a distinctive historical configuration. In a time and place that was by most measures still dominated by such "premodern" features as absolutism, alchemy, and astrology, the epicenter for this new configuration can plausibly be located in the first modern nation-state, the Dutch Republic, and within it, Amsterdam as the principal hub.

In setting this primal scene with Descartes and Galileo, I am obviously moving within a retroactively constituted history. At the time, nobody could have foreseen the power that history would bestow on these characters as markers of a new and (for us today) self-evident beginning to an ostensibly linear history of progression, forged out of a chaotic play of multiple forces.

In logical terms, the kind of identity loop that makes such a distinct origin marker possible is governed by the most powerful metaphysical principle of Western logic, what G. W. F. Leibniz named the principle of contradiction, sometimes referred to as the principle of noncontradiction—namely, that something cannot be and not be at the same time. A is A and cannot simultaneously be non-A. Heidegger called this the *principle of identity*. A = A, its identity, its sameness, is A. Today, it is a commonplace assertion that modern philosophy begins with Descartes, and this repeated determination ensures his right to a place in history. As Heidegger shows, the principle of identity governs how something comes to be some thing, discernible and distinct—that is, the representation of Being. At first glance, A = A is a perfect tautology. It says that A is the same as A and that every A is everywhere the same, in the sense of the Greek *to auton*, from which the Latin *idem*, for "identity," derives. A = A bespeaks self-sameness. But more fundamentally, Heidegger says, A = A also means that "every A is itself the same with itself. Sameness implies the relation of 'with,' that is, a mediation, a connection, a synthesis: the unification into a unity. This is why throughout the history of Western thought identity appears as unity."[6]

Whenever we speak of a thing or a being, any appearance of identifiable, self-bounded individuality, we speak using the principle of identity as our highest key. The principle suggests itself as a naturally given premise for comprehending the world, by thinking in terms of units and unity. For insofar as the principle expresses what something is by virtue of what it is to itself, it also speaks to the conditions for how a being comes into being as distinctive and clearly differentiated. In this sense, writes Heidegger, the principle of identity appears in Western thought as both an ontological ground and a theological whole:

> What the principle of identity, heard in its fundamental key, states is exactly what the whole of Western European thinking has in mind—and that is: the unity of identity forms a basic characteristic in the Being of beings. Everywhere, wherever and however we are related to beings of every kind, we find identity making its claim on us. If this claim were not made, beings could never appear in their Being. Accordingly, there would then also not be any science. For if science could not be sure in advance of the identity of its object in each case, it could not be what it is.[7]

The self and the relation of the self to itself—that is, what A is with respect to A—is therefore properly a "belonging together." What A is said to be belongs together with what A is. But in what sense? The limit of the scientific understanding of identity, Heidegger argues, is that it effectively effaces the mediating relation to itself—it excludes its own middle. "Belonging together" is determined by the word "together" so as to assure its unity. To belong in this sense means "to be assigned and placed in the order of a 'together,' established in the unity of a manifold, combined into the unity of a system, mediated by the unifying center of an authoritative synthesis." However, belonging together has a potentially inverse meaning that is effectively concealed to representational thinking. "How would it be if, instead of tenaciously representing merely a coordination of the two in order to produce their unity, we were for once to note whether and how a belonging to one another first of all is at stake in this 'together'?"[8]

What is lost to the modern scientific mind, Heidegger argues, is precisely how something connects to everything rather than how something is a single thing for itself. This chapter is about this pivotal logical difference, like two different dimensions of thinking, between Descartes's logic of the *cogito* versus Spinoza's logic of connection. As Heidegger explains, the principle of identity understood as the absolute first principle of Western logic enables the physicist today to encounter the world first and foremost as an agglomerate of already differentiated particles, before considering their connection: a relation thought in terms of its particles rather than the other way around. As I wrote in the previous chapter, to begin with the idea of identifiable particles is above all a necessary presupposition in order to do the work, even as the research increasingly points toward the other dimension, in which no particular identity appears stable but rather relational to the experiment itself. Perhaps this is an example of Heidegger's maxim that every revealing is simultaneously a concealing—the self-evident truth of A = A hides its inverse dimension.

The inversion of identity can be highlighted through a second fundamental principle of thought—the *principle of reason* (or what Leibniz called the principle of sufficient reason). The principle of reason says that "nothing happens without a reason that one can always render as to why the matter has run its course this way rather than that."[9] A principle of thought as much as a principle of causality, it becomes a metaphysical determination, or constraint, for physical explanations. In order to understand how something comes into being, we have to inquire into a relationship. How did A become A? Operating together, the principle of identity and the principle of reason constitute the basic matrix of modern metaphysics. As I will show in this chapter, these two principles can relate in different ways, depending on which of the two reigns supreme.

Michel Serres explains the relationship between these two principles thus:

> If we had only the principle of identity, we would be mute, motionless, passive, and the world would have no existence: nothing new under the sun of sameness. We call it the principle of reason that there exists something rather than nothing. From which it follows that the world is present, that we work here and that we speak. Now this principle is never explained or taken up except in terms of its substantives; the thing, being and nothingness, the void. For it says: *exist rather than*. Which is almost a pleonasm, since existence denotes a stability, plus a deviation from the fixed position. To *exist rather than* is to be in deviation from equilibrium. Exist rather. And the principle of reason is, strictly speaking, a theorem of statics.[10]

In other words, rather than identity appearing out of itself, there is always already something deviating from this unity, a given premise. The principle of reason gestures toward this positive existence. Leibniz's Latin formulation, *Nihil sine ratione*, nothing without reason, could therefore be inversely restated along Serres's lines as *Semper sic*—always something on the condition of something else, always this rather than that, always a given—never a unit only for itself.

This positive given premise for something actually existing, which we may not be able to know but for which there is necessarily a reason, is what I will call the *autological*. This is the self-positing logic of that which gives itself in and for itself, but which is not itself given as a self—or put differently, what is not already given under the principle of identity as a clearly differentiated thing. In the history of thought, the autological occurs in myriad configurations. In physics it becomes primarily expressed as force, gravity, which becomes the foundational problem of dynamics, and later as energy. In Scholastic philosophy, the vernacular of the seventeenth century, the autological is principally substance, the *causa sui*, the self-causing cause—and with Spinoza's *Ethica* the logic is axiomatic as simultaneously substance, God, and Nature. In theology, the autological is often expressed as the logic of the soul, that which animates us as thinking beings.

However, to modern science and epistemology this unquantifiable autological dimension has been treated with suspicion, and since the twentieth century the very idea behind it has largely disappeared from view. When scientific cosmologists today conceive of the universe, they first think of it mathematically through a void, and in this manner they are Galileo's descendants. But, as I will try to demonstrate in this chapter, when the premise for thought is excluded from thinking and our conception of the world order is turned inside out from our affective experience, strange and gaping problems follow that not even the most complicated mathematics can answer.

GALILEO'S VOID

When we follow the principle of reason to its limit, metaphysics becomes the vanishing point between believing and knowing, theology and science. The mid-seventeenth century constituted a rare historical convergence between physics and metaphysics under the twin signs of natural philosophy and natural theology. As Stephen Gaukroger puts it in his study of the emergence of modern scientific culture:

> A good part of the distinctive success at the level of legitimation and consolidation of the scientific enterprise in the early-modern West derives not from any separation of religion and natural philosophy, but rather from the fact that natural philosophy could be accommodated to projects in natural theology: what made natural philosophy attractive to so many in the seventeenth and eighteenth centuries were the prospects it offered for the renewal of natural theology. Far from science breaking free of religion in the early-modern era, its consolidation depended crucially on religion being

in the driving seat: Christianity took over natural philosophy in the seventeenth century, setting its agenda and projecting it forward in a way quite different from that of any other scientific culture, and in the end establishing it as something in part constructed in the image of religion.[11]

With natural philosophy and theology as the axis of modern scientific culture, the autological dimension was considered a given, whether it was understood as natural force on the one hand or as divine presence on the other. Physics and metaphysics were considered mutually indispensable in making sense of this same positive reality, which soon became the central problem in a new idea that emerges in this era, which became known as the *universe*.

In *Philosophiae Naturalis Principia Mathematica* (1687), which within a century of its publication had become the touchstone of modern physics, Isaac Newton offers a new mathematical world-picture: in one and the same logical operation, one and the same configuration, it yields a complete and unified realm of quantitative explanation. From Newton's *Principia* onward, through the conceptual extension and simplification that Einstein offers it under general relativity, the universe becomes a self-evident name and concept for the unified totality of the cosmos within which humans find themselves.

Newton's universe was predicated on "Rational Mechanics," which, as he put it in his preface,

> will be the science of motions resulting from any forces whatsoever, and of the forces required to produce any motions, accurately proposed and demonstrated. . . . And therefore we offer this work as mathematical principles of philosophy. For all the difficulty of philosophy seems to consist in this—from the phenomena of motions to investigate the forces of Nature, and then from these forces to demonstrate the other phenomena.[12]

Traditionally, mechanics consisted of three areas: *kinematics* deals with bodies already in motion, *statics* with bodies in a state of equilibrium, and *dynamics* with forces responsible for motion. Dynamics was, as Gaukroger puts it, "the ultimate prize of 17th century physics" because it had to, like Newton's work, encompass both a theory of motion and of forces.[13] But in order to accomplish this in quantitative fashion, the development of dynamics had to encompass two other areas that would turn out to be mutually exclusive. Statics, well developed through antiquity, deals with forces but not with motion. Kinematics, on the other hand, deals with motion but not with forces. To natural philosophers seeking to pursue a complete model for dynamics, statics and kinematics therefore offered two different routes to the same ostensible goal. But each route implies different metaphysics, and thus different realities. As it would turn out, only one of these routes allowed for the quantification necessary to constitute a mathematical universe of the kind we know today. And it was Galileo, after decades of working with models from statics that proved unsuccessful, who eventually found a way through kinematics, which became the foundation for Newton's work.

Galileo's trailblazing work owes much to a new institutional context that both enabled and compelled him to frame his problems in novel ways.[14] As the historian Mario Biagioli points out, the revolutionary break that Galileo, as a courtier in the aristocracy of Italian city-states, makes with Scholastic thought is also a break with a Scholastic institution. In Galileo's extended networks of patronage, social status provided more authority than what we now think of as scientific credibility. As Biagioli shows, Galileo's success as a courtier was contingent on his ability to mediate interest within a newly emergent domain for natural philosophy between the traditional, Scholastic order of knowledge and the later order emerging through scientific academies.

As both the enactor of the experimental observation and the inventor of the conditions for the mathematical framework within which this observation can be explained and legitimated, Galileo comes to play a crucial double role in the making of scientific history: Galileo the experimenter and Galileo the mathematizer. In physics, Galileo's mathematical legacy comes from his principle of invariance, which we find in both the Newtonian and Einsteinian universe, the idea that the laws of physics are the same in all "inertial frames." Experimentally, Galileo's legacy hinges on his decisive demonstrations, first of the telescope, which opened up a new perspective, but shortly thereafter of the inclined plane, which would prove consequential to the development of dynamics.

The inclined plane is an abstract representation of space and time through which Galileo wanted to demonstrate his theory of motion. As discussed in chapter 1, the constitutive role of the experiment in modern science is its ability to play on a double register—as Isabelle Stengers puts it, it makes the phenomenon "speak" in order to "silence" the rivals. And it is no coincidence that Galileo's analytical reduction of motion into separable elements happens to be precisely what his device is designed to demonstrate.[15] In a close reading of Galileo's device as the proto-experiment of modern science, Stengers points out that, contrary to the conventional understanding of the experiment as a positive demonstration of truth, the inclined plane, as a laboratory rendition of the world, first and foremost establishes a negative truth that only retroactively comes to appear as a positive statement about nature. Because the apparatus allows its author, Galileo, to withdraw and instead let the premeditated motion testify in his place, it appears as though nature is made to "speak" directly through the experiment. But the logic of the experiment is more circuitous. Stengers continues:

> The "law of motion" is not linked to observation but is relative to an order of created "fact," to an artifact of the laboratory. But this artifact has a singularity: the apparatus that creates it is also able, certainly not to explain why motion lets itself be characterized in this way, but to counter any other characterization.[16]

Galileo's primary rival fiction was Aristotelian physics, which he defeated by reconfiguring the role of the experiment and allowing for physical problems to be explained mathematically.[17]

In Galileo's curious 1638 text, the *Discorsi*, the new rationale for kinematics is laid out as an ongoing dialogue between three characters—Salviati (the interlocutor), Sagredo (the skeptic), and Simplicio (the Aristotelian)—referring and reacting to statements by the Author (Galileo). In the Aristotelian view, prevalent at the time, motion is itself an irreducible physical reality underlying time as a mental abstraction. Galileo, however, treats motion purely as a local change of spatial location in time. With help from the inclined plane, he asserts that motion can be divided into three independent forms. First, uniform motion "is defined by and conceived through equal times and equal spaces (thus we call a motion uniform when equal distances are traversed during equal time-intervals)." Uniform motion is similar to what Newton would call inertia, straight rectilinear movement along a horizontal plane, like a billiard ball on a snooker table. Second, naturally accelerated motion is a motion "uniformly accelerated . . . starting from rest, it acquires, during equal time-intervals, equal increments of speed."[18] This is, in other words, the case of free fall or vertical motion toward the ground. Third, projectile motion, or projection, is, Galileo argues, a compound of uniform and accelerated motion, both of which are demonstrable through the inclined plane. His novel thesis is that "a projectile which is carried by a uniform horizontal motion compounded with a naturally accelerated vertical motion describes a path which is a semi-parabola," a perfectly sloping geometrical curvature.

Given their turn, Sagredo and Simplicio counter with three objections to the Author's geometrical argument. First, Sagredo points out that the semi-parabola, which is in theory perpendicular to a horizontal surface, cannot account for the tendency of a falling body to fall toward the center of the Earth, which the geometrical abstraction could never reach. In post-Newtonian language, gravity would in some way intervene on the perfect geometrical path by drawing it to Earth, and "the path of the projectile must transform itself into some other curve very different from the parabola." Second, as Simplicio notes, the Author supposes "the horizontal plane, which slopes neither up nor down, to be represented by a straight line as if each point on this line were equally distant from the center, which is not the case; for as one starts from the middle (of the line) and goes toward either end, he departs farther and farther from the center of the earth. . . . Whence it follows that the motion cannot remain uniform . . . but must continually diminish." In other words, Galileo's demonstration cannot take account of the sphere of the Earth. Finally, Simplicio adds: "I do not see how it is possible to avoid the resistance of the medium which must destroy the uniformity of the horizontal motion and change the law of acceleration of falling bodies." That is, Galileo's demonstrations are valid insofar as the world is a flat, forceless, friction-free space—a void—which is clearly absurd. The first two objections both concern gravitation, insofar as the curved shape of the Earth is directly implicated in the tendency for bodies to fall to its center. The latter objection concerns force in a more immediate sense. Faced with the critique, Salviati, on behalf of the Author, admits "that these conclusions proved in the abstract will be different when applied in the concrete and will be fallacious to this extent." Nevertheless, he counters—and herein lies the essential invocation that makes the Newtonian universe possible—"in order to handle this matter in a scientific way, it is necessary to cut loose from these difficulties; and having discovered and demonstrated the theorems, in the case of no resistance, to use them and apply them with such limitations as experience will teach."[19]

In short, the kinematic approach to dynamics lies in the wholesale removal of the medium in which bodies move. Galileo is the first to introduce this fundamental feature of contemporary physics: first, the world is removed, and second, it is reintroduced piecemeal. What is at stake in such an "illogical" maneuver (as it appeared at the time) is the ability to mathematize physics. In the text, Sagredo and Simplicio are appeased by this concession and allow Salviati's interlocution with the Author's subsequent theorems and demonstrations to continue. In order to be scientific, we have to act *as if* the world is (in this case) an absurd void. Thus, a whole new set of inventions become possible.

First of all, Galileo's kinematics privileges the fundamental relativity of motion, wherein rest and uniform motion (inertia) are identified as equivalent states. In kinematics, differences of motion are never an absolute difference between moving and not-moving but rather relative differences between moving-less and moving-more. This can be used to express ratios, or relations of change, for speed, momentum, weight, and so on. Galileo becomes the first to render motion in terms of a state, using the principle of identity to turn the autological reality of motion into a hypostasized expression of relative differences.

Second, by arguing through the (absurd) condition of a void, Galileo is immediately able to generalize his relative problem in a transformative way. By way of an extended principle of identity, the universal similitude of all bodies, regardless of their mediated situation, is first of all established, and under this "Archimedean point," the world as it exists, as a given, can be treated as simple mathematical differentiations. No longer is it a question, as in the Aristotelian doctrine, of different bodies under different conditions, but rather any bodies with any weight or any speed, since all their worldly imprecisions and impediments have been first stripped and subsequently redressed. In this way, the stubborn discrepancy between an idea (of a movement in space) and its actually existing conditions (forces in a physical place) is circumvented. Enframed as a constant, Galileo's claim becomes valid in all "inertia frames" precisely on account of the abstracted void, which provides a stable referent that enables the quantification of movement.

Thus, Galileo's invention is a matter of carefully reconstituting a problem of motion in order to circumscribe the force of the given and remove the mediation that makes mathematization problematic. In the *Discorsi*, Galileo takes readers through a painstaking reconstitution of the relative weight of air to bodies—the medium his generalization needed to remove—and through these sequential experiments, piece by piece reconstituting what was initially removed, Galileo is able to make a plausible claim to having realistically established, or rather reestablished, the very situation that is by definition impossible to submit to experience. Whereas the autological medium was once considered constitutive of the physical problem as such, it has now been redefined exclusively in terms of resistance to an idealized motion. In turn, the explanatory structure of physics is transformed: rather than beginning from the mediating force in which the natural philosophers find themselves, the "new science" of Galileo begins with the circumscription of the world. Galileo's invention is not merely recourse to an experimental device, nor simply the a priori hypothetical generalization to which this device is put, but rather a mutual construction: the inclined plane enables Galileo's analytic breakdown of motion into constituent parts, which it in turn demonstrates through the synthetic reconstruction of the circumscribed reality. If this double logic appears bewildering, it is perhaps because it takes considerable metaphysical contortion to make the world disappear.[20]

In short, after Galileo, the autological world becomes the excluded middle.

DESCARTES'S VORTEX

Until Galileo's kinematics became the basis for Newton's mathematical universe, the most influential cosmology was developed by Descartes, the preeminent natural philosopher well into the eighteenth century. Descartes attempted to arrive at dynamics through a different route—statics. And though it appears in retrospective history as a detour compared to Galileo's path, it nonetheless turned out to be consequential, for Descartes's struggles would cause him to provide the new scientific universe with a significant metaphysical legitimation.

Descartes's conception of the cosmos agreed with Galileo on the heretical question of heliocentrism. As a natural philosopher, Descartes was also aligned with Galileo in the general project of forging a mathematical physics that would overturn the traditional hierarchy of the Aristotelian sciences. Certainly Galileo would not object to Descartes's stance that "the only principles which I accept, or require, in physics are those of geometry and pure mathematics; these principles explain all natural phenomena, and enable us to provide quite certain demonstrations regarding them."[21]

Nevertheless, when Descartes went back to his bookseller in 1638 to pick up Galileo's next book, the *Discorsi*, his review of it was harsh. He dismissed the Italian philosopher's kinematic rationale out of hand. For Descartes, who had spent much of the two previous decades grappling with dynamics, Galileo's work was problematic for two fundamental reasons. First, it provides no account of causality—it deals with motion in itself but not with the forces that cause motion. That is, logically speaking, it submits motion to the principle of identity without using the principle of reason. Second, Descartes considered the void to be a logical and physical absurdity, because there exists no vacuum in nature. For Descartes, the presence of force is a given condition that precludes thinking in terms of its absence; that is, in the Cartesian world-picture, the autological given of the world has to be an included middle.

In his book *The World*, Descartes first tries to argue that the phenomenal qualities of light can be explained in terms of motion, and that the generalized phenomenon of motion extends to all that we know as Nature, that is, the ever-changing realm of physics. Henceforth, he is dealing with a pure mechanism with three laws of nature that, contrary to Galileo, are modeled on hydrostatics. In statics, the paradigmatic instrument is the *equilibrium*—the scale or beam balance that will incline or decline depending on the weight distributed on either side.[22] What statics most essentially measures is deviation from a constructed equilibrium, and this procedure has implications for the questions one can ask. Logically, statics first defines a rest position, a degree zero, then a movement as an absolute difference from this initial position. The principle of identity and the principle of reason here operate in tandem. On the one hand, through the rest state of balance, identity is defined, from which we may ask about the reason or cause of the motion. Thus through the principle of identity, we reinforce the classical, Aristotelian divide between motion and rest as absolute categories—at the pivot, either the scale is moving or it's not, one or zero, A or not A. On the other hand, when we employ the principle of reason on this difference between rest and motion, the answer is not motion itself but rather the limit condition of motion, the point at which motion begins. For Descartes, this limit condition will be understood as "tendency to motion." Force enters the picture as we increase or decrease the weight and the scale tends in either one or another direction. When we ask about why the scale moves or not, or why it moves in this case but not that, we are asking about force, about what is directly bearing, through its presence, on the identified difference. From the point of equilibrium, we are inquiring into the causes for that which appears as the autological given.

Descartes's first law of nature posits that "each particular part of matter always continues in the same state unless collision with others forces it to change its state." This looks similar to Newton's law of inertia, but it is actually different, for in Newton's case inertia is derived from the absence of the force that Descartes is in fact attempting to include in his description. This becomes clearer with the second law, which resembles a law of conservation: "When one of these bodies pushes another it cannot give the other any motion except by losing as much of its own motion at the same time; nor can it take away any of the other's motion unless its own is increased by the same amount."

Descartes is trying to account for a universe in which the sum total of motions never changes. The differentiation of movement within this plenum, in which all parts of matter move and are moved by others, is then fully realized with the third law: "When a body is moving, even if its motion most often takes place along a curved line . . . it can never make any movement that is not in some way circular. Nevertheless, each of its parts individually tends always to continue moving along a straight line. And so the action of these parts, that is, the inclination they have to move, is different from their motion."[23]

This distinction will become more explicit in Descartes's later work, *Principia*: whereas the tendency toward motion is rectilinear, actual motion is always circular. That is, we are dealing with two mutually constitutive forces or tendencies derived by way of difference: one that strives to deviate from being kept in place by the other. Nature, according to Descartes, fundamentally operates in a circularity, whose constant change emerges in, as it were, a deviation from itself. Herein lies the logical flip needed to constitute a Newtonian universe. Whereas Descartes attempts to explain how straight motion is possible within an autological universe of force which operates with circular motion, Newton inverts the problem and asks how force intervenes to turn kinematically straight movements back into the circular motion of the cosmos.

Newton's retroactive constitution of the problem, which we recognize from Galileo, is pivotal to making the universe quantifiable. But it comes with a strange discrepancy: the dimension of reality that is most immediately present, the force of gravity, becomes for Newton an external and mysterious force acting at a distance, left as a circumscribed constant that makes calculations work but which cannot be explained. Although Einstein's general relativity theory later integrates gravity into the idea of the universe, this merely shifts the problem to a different level of explanation, because gravity becomes the limit condition between relativity and quantum physics and, as I argued in the previous chapter, still constitutes the stumbling block for all efforts at achieving mathematical universality. Incidentally, this is also the case with light, which neither Newton nor his successors can account for other than in exceptional terms. The autologically given phenomena of gravity and light constitute the limit conditions of Newton's kinematically derived dynamics and stand for the problematic kernel of modern physics to this day. In this sense, the modern cosmological framework is predicated on their exclusion, but their subsequent inclusion causes the general instability of the model.

Descartes's account of light and gravity, on the other hand, makes immanent sense to his model, because he has already defined matter autologically. The Cartesian universe is fundamentally mediated. According to his world-picture, we live in a fluid matter of planets that resembles a chunky soup, an infinitely extended substance differentiated solely in terms of the different motions of its parts. Through this extensive matter, light is transmitted as waves. Descartes's considerable contributions to optics and the first phase of microscopic technological innovation owes much to his conception of light as a fundamental continuity, which would influence scientists like Christiaan Huygens and subsequent theories of light. As for gravity, Descartes does not conceive it as a separable or abstract force but rather as a direct consequence of the pressure caused by the swirling nature of matter:

> I want you to consider what the weight of this Earth is, that is, what the force is that unites all its parts and makes them all tend toward the center, each more or less according to the extent of its size and solidity. This force is nothing but, and consists in nothing but, the parts of the small heaven which surround it turning much faster than its own parts about its center, and tending to move away with greater force from its center, and as a result pushing the parts of the Earth back toward its center.[24]

The hydrostatic influence on Descartes can here also be understood as an extension from the elemental quality of water, insofar as his universe is constituted by omnipervasive fluid matter whose flux is the analogical concept for the cosmic order. Descartes's essential cosmological idea is perhaps most eloquently expressed in the *Principia*:

> The whole of the celestial matter in which the planets are located turns continuously like a vortex with the sun at its center. . . . The parts of the vortex which are nearer the sun move more swiftly

than the more distant parts, and . . . all the planets (including the earth) always stay surrounded by the same parts of celestial matter. This single supposition enables us to understand all the observed movements of the planets with great ease, without invoking any machinery. In a river there are various places where the water twists around on itself and forms a whirlpool. If there is flotsam on the water we see it carried around with the whirlpool, and in some cases we see it also rotating about its own center; further, the bits which are nearer the center of the whirlpool complete a revolution more quickly; and finally, although such flotsam always has a circular motion, it scarcely ever describes a perfect circle but undergoes some longitudinal and latitudinal deviations. We can without any difficulty imagine all this happening in the same way in the case of the planets, and this single account explains all the planetary movements that we observe. (IIIP30)

The solar system, then, as flotsam in a whirlpool. Descartes's intuitive account of planetary movements generated much interest among his contemporary natural philosophers. The vortex theory of planetary motion, conceived by Descartes and developed by his disciples, became the dominant cosmology in the mid-seventeenth century.

Nevertheless, despite his many achievements in quantifying natural phenomena in the realm of practical mathematics, Descartes could provide no such quantitative extension to his natural philosophical system as a whole. The dream of a *mathesis universalis* on his philosophical terms, as an alignment of the three theoretical sciences of Aristotle—physics, mathematics, and metaphysics—could not be actualized.[25] Descartes could not think in terms of inertia of the Galilean kind, because it was wholly alien to his metaphysics. Indeed, these two different modes of mechanical thinking appear mutually exclusive: Descartes's "failure" to provide a thoroughly quantitative system hinges on the very same qualitative (and unquantifiable) idea of infinite extension, which made his system so cogent in the first place.

As the kinematic case of Galileo indicates, in order to quantify something we need both a discontinuity (an identifiable difference) and a stable reference point. Descartes's identification of extension as the essence of physical substance, infinitely divisible into different parts, was designed to allow for mathematical treatment. However, the only stable reference in Descartes's early physics was ambiguously situated outside Nature: namely, God. However God is defined, he makes for a poor mathematical constant. Operationally, we require something that exists within Nature, not outside it. After the church condemned Galileo, Descartes was hard at work on this problem. He held back publication of *The World,* and his idea of a world-picture therefore first appeared, arranged more systematically, in the 1644 publication of the *Principia.*

Crucially, the physics of the *Principia* is preceded by a first part on "the principles of human knowledge," explaining the relation between the human mind, nature, and God in a manner that for Descartes reconciles a mathematical dynamics of the universe with the church's ban on heliocentrism. As he puts it in a principle that determines the fundamental relativity of positions between bodies in the world, "if we suppose that there are no . . . genuinely fixed points to be found in the universe (a supposition which will be shown below to be demonstrable) we shall conclude that nothing has a permanent place, except as determined by our thought" (IIP13). Thus, a novel idea emerges: thought itself as the constitutive exception. In the course of the first eight principles of the *Principia*, Descartes establishes what now becomes an attempt, in Arendt's phrase, "to move the Archimedean point into man himself, to choose as ultimate point of reference the pattern of the human mind itself, which assures itself of reality and certainty within a framework of mathematical formulas which are its own products."[26] The discrepancy between thinking and being, in other words, is now internalized, into the *cogito.*

How does Descartes arrive at this certainty? His crucial first principle of human knowledge is that "the seeker after truth must, once in the course of his life, doubt everything, as far as possible"—an attempt at "freeing ourselves" from our preconceived opinions (IP1). Similar to Galileo's method, the world first needs to be removed. In Descartes's metaphysics, then, if not his physics, the autological once again becomes the excluded middle. Descartes circumscribes the world by arguing, in his second principle, that whatever can be doubted should be considered false. The implication is that when thought is considered independent of the world, there is a fundamental bifurcation of things, truth on the one hand and falsity on the other. In the pervasive analogy throughout Descartes's work, this corresponds to the two autonomous dimensions of light and darkness. The third, fourth, and fifth principles function as elaborations and qualifications for this procedure, arguing that while such doubting should not be attempted in the course of "ordinary life," in which we require our senses, it is precisely the senses that can deceive us and therefore warrant removal.

Descartes argues that this experiment of thought can be replicated by anyone. What it first requires is analogical to the procedure of statics—a definition of rest state (everything in doubt), from which we can employ the principle of reason to infer the cause of our doubting. However, for Descartes this cause has an identity. Presumably by the fact that we can choose to undertake such an experiment, Descartes deduces an important sixth principle: "We have free will." The "fact" of free will, Descartes argues, is what enables our doubting, for we can thus freely withhold our assent in potentially false matters. Given this supposed freedom, and given the removal of the senses, through which a fundamental distinction between the true and the false can be made, Descartes famously concludes in his seventh principle: "It is not possible for us to doubt that we exist while we are doubting; and this is the first thing we come to know when we philosophize in an orderly way." To philosophize in an orderly way means, for Descartes, to first remove the mediation of our existence and, as a corollary, submit to the principle of identity. "For it is a contradiction," Descartes warns, "to suppose that what thinks does not, at the very same time when it is thinking, exist" (IP7). In other words, the essence of the *cogito* ("the natural light of our soul") is that whatever is doing the thinking must be, as per the principle of identity, a thinking thing. Descartes's conclusion lies in the eighth principle: "In this way we discover the distinction between soul and body, or between a thinking thing and a corporeal thing." Here is his explication of the principle:

For if we, who are supposing that everything which is distinct from us is false, examine what we are, we see very clearly that neither extension nor shape nor local motion, nor anything of this kind which is attributable to a body, belongs to our nature, but that thought alone belongs to it. So our knowledge of our thought is prior to, and more certain than, our knowledge of any corporeal thing. (IP8)

If we examine it carefully, we find that Descartes's reasoning is already implied in the supposition that we can remove our minds from the world through an absolute division between truth and falsity. Indeed, in the French version of *Principia* the supposition reads, "if we who are now thinking that there is nothing outside of our thought which truly is or exists . . ." In other words, he derives his conclusion from his premise, and both are constituted by appeal to the principle of identity. Having clearly set up this new equilibrium between two identities, mind and body, Descartes then turns to the principle of reason to argue for the prior certainty of the mind, its "natural light" of truth. In the eleventh principle he says:

We should notice something very well known by the natural light: nothingness possesses no attributes or qualities. It follows that, wherever we find some attributes or qualities, there is necessarily some thing or substance to be found for them to belong to; and the more attributes we discover in the same thing or substance, the clearer is our knowledge of that substance. (IP11)

Nothingness has nothing, so there must be something. Nothing without reason—nothing has no reason—so insofar as we find something, this something clearly belongs to an identified thing. Throughout Descartes's text there is at work a deft logical play in which the principle of reason is always employed in the service of an ultimate identity. And this is how Descartes approaches the problem of substance itself, the autological given. For if the mind can come to know itself as a thinking thing, distinguished from the world as a corporeal thing, how do we know that this certainty is not merely a mental delusion? In the thirteenth principle he concedes that knowledge of things "depends on the knowledge of God": "the possession of certain knowledge will not be possible until it has come to know the author of its being." But God can be ascertained, he argues in the fourteenth principle, from the fact that in the mind "there is one idea—the idea of a supremely intelligent, supremely powerful and supremely perfect being—which stands out from all the others." This is the idea of necessary existence, that which must exist in order for anything else to exist. What is this necessary existence? Logically, it is a postulate of the principle of reason—nothing is without cause—as determined by the principle of identity. For Descartes, God as this necessary reason for existence is a being, or a thing, the Thing of all things. And insofar as God is the necessary reason both for our thought and for that which we distinguish as matter, this Thing of things is the constitutive cause of our knowledge of the world.

Thus, the frantic search for a legitimate grounding of his natural philosophy takes Descartes in two simultaneous directions—between mind and God on the one hand and between God and matter on the other. That is, whereas Galileo and Newton could predicate the project of a mathematical physics on the invention of a void, Descartes tries to conceive his project metaphysically, by recourse to theology on the one hand and epistemology on the other, in a mutually reinforcing framework. The stable referent for Descartes's autological universe, which makes his world-picture non-heretical, is constituted by aligning God and the human mind in such a way that certain, mathematical knowledge of Nature becomes possible.

UNIVOCAL VERSUS UNIVERSAL

However, in order for theology and epistemology to reinforce one another into a complete mathematical universe, the relation between God and Nature must be of a certain configuration. In the *Principia*, as he moves from the individual mind in doubt to the conditions for how the mind can know with certainty, Descartes eventually arrives at the autological concept of substance:

> By *substance* we can understand nothing other than a thing which exists in such a way as to depend on no other thing for its existence. And there is only one substance which can be understood to depend on no other thing whatsoever, namely God. In the case of all other substances [i.e., mind and body], we perceive that they

can exist only with the help of God's concurrence. Hence the term "substance" does not apply *univocally*, as they say in the Schools, to God and to other things; that is, there is no distinctly intelligible meaning of the term which is common to God and his creatures. (IP51)

In the Scholastic vernacular, the *vocal* is logically distinct from the *versal*, in the same way that utterance (the sound we make when we speak) and meaning (the words we use) constitute two different dimensions of the same reality. The word *vocal* signifies an initiation of sound. *Versal*, a Latin word associated with "text," stands etymologically for "turning." The word *versus* still carries this original sense of "turned" (past participle of *vertere*) in the everyday sense of "against"—as in, Galileo "turned against" Descartes. In the difference between the vocal and the versal we can recognize the distinction between a logic of a given and a logic of turning, between a primary presence and a secondary folding.[27] The paradox is that only through the logic of turning (words) could we express our understanding of that which is given (sound)—yet this given is the condition for that which is turned into thought and expression. This relation of thought to being, turning to initiation, is like a constantly twirling dance between the principles of identity and reason, which must both move with each other. But the question is, which principle leads in this dance?

For Descartes, the relation between God and Nature is, as he writes, not *univocal* but rather *equivocal*. In this context, *equi* signifies a relation of difference, of more than one, as in the case of the statical equilibrium under the principle of reason, in which the balance state (equal weight) is caused by something other than itself. Equivocity means that nothing can be said of Nature or of a creature in Nature that can be said of God in the same sense; that is, God is distinct from Nature, not one with it. Following the biblical story of creation, equivocity implies the absolute distinction between creating and created substance: God is the maker of the universe, not the universe itself. So even though mind and body for Descartes are distinct attributes, they are unified in their ultimate creating cause, namely, God.

A universal mathematical physics is possible, Descartes argues, because the mind is in such a privileged position that it can, insofar as it "reasons in an orderly way," clearly perceive the nature of matter. The mind, in other words, is like a calm eye in the storm, a fixed Archimedean point within the relentless vortex of Nature that can grasp the movement surrounding it without itself being carried away. This is only possible on condition of an equivocal relation between God and Nature. In other words, *equivocity* is the necessary condition for *universality*. For Descartes, the concept of the universe is a direct logical consequence of this configuration of God and Nature. This means the scientific universe follows from the logic of the Christian doctrine of creation: the creator-being separated from created being.[28] Only when the principle of reason is thus submitted to the ultimate identity—God—can a universe be born.

Although the Galilean-Newtonian route to dynamics veers methodologically from Descartes's path, they are nonetheless close kin in the emergence of the modern scientific universe, as they both overturned Aristotelian natural philosophy. In the subsequent historical divergence of natural philosophy into philosophy on the one hand and science on the other, the Cartesian and the Galilean-Newtonian inventions would turn out to be mutually reinforcing legitimations. Read against each other, Newton explains the mathematization of the physical universe, while Descartes explains the universal grounding of this mathematization in thought. Both accounts of universality rely on God as the equivocal author of the universe. And both accounts rely on the constitutive circumscription of the world as given. Whereas Galileo excludes mediation in physics through the void, Descartes excludes it in metaphysics through the *cogito*.

In other words, both Galileo and Descartes constitute a universe that turns the autological inside out and becomes, in Hegel's phrase, "the inverted world." But what kind of metaphysics would be possible without such a radical inversion?

SPINOZA'S VOICE

As Descartes was twisting and turning his thoughts on his way back from the bookseller in Amsterdam, desperately looking for a way to square his cosmology with the teachings of the Catholic Church, he would have passed by the Jewish quarter of the city. Here, among

Spanish and Portuguese Jews who had fled forced conversion and inquisition by the same religious authorities to which Descartes wanted to defer, lived a child called Baruch Spinoza. He would eventually be excommunicated from his own community and set out on a life in search of his own philosophical answers. Spinoza's philosophical role was to revert Descartes's world-picture into a fully rational system, consistent with Descartes's own desire for an orderly philosophy, that could fully account for the autological dimension of the world.

However, Spinoza's idea came at a hefty cost, as it was deemed heretical not only to Jews and Christians but also to the modern science that was beginning to emerge. Although Spinoza is rarely considered in terms of his contributions to physics, as the preeminent metaphysician of the early modern era, a brief overview of his world-picture can shed light on the alternative perspective from Galileo, Descartes, and Newton.

While Descartes fashioned himself an optical theorist, Spinoza occupied himself as a lens grinder, and perhaps his practical knowledge made his perspective different. In either case, their philosophical difference can be distilled in a plain metaphysical insight on the nature of light itself. For both thinkers, the presence of light bespeaks God most clearly and distinctly. But in Spinoza's *Ethica*, published posthumously in 1677, it does so in a very different sense than Descartes's ontological bifurcation of light and darkness, truth and falsity. Writes Spinoza: "As the light makes both itself and the darkness plain, so truth is the standard both of itself and of the false" (IIP43S).[29] Darkness is not the opposite of light but rather its limit condition. Falseness is not its own state but the limit condition of truth. Thus, in one image, Descartes's world premise is turned inside out, his ontological bifurcation turned into an inherent relation. In the metaphysics of light, two different world configurations stand exposed, like positive and negative images of each other.

Similarly, Spinoza's *Ethica* reads like an inversion of Descartes's *Principia*, in structure as well as in idea. Whereas Descartes begins with the individual thinker and then reasons his way toward God and Nature, Spinoza begins with God and reasons his way to the individual thinker in Nature. And whereas Descartes's individual is a lonesome doubter in self-imposed isolation from the world, Spinoza's individual finds itself constantly affecting and being affected by others. We find in Spinoza's philosophy the same dance between the principles of identity and reason, but for Spinoza it is reason that leads, and this makes for a radically different dynamic.

Crucially, Spinoza's starting point is that substance, the autological given of the world, is fundamentally active. All things are made of substance and exist through their active "essence," which is their *conatus*, or striving quality: that "each thing strives to persevere in its being is nothing but the actual essence of the thing" (IIIP7). Because the principle of reason reigns supreme over identity, the essence of a thing is not itself a thing. Rather, the essence is this striving, the tendency to persist. The thing may be considered passive in that it is acted upon and determined by other things, but it is active insofar as it maintains its determination and tries to exist in this way rather than that. The striving tendency thus expresses an autological dimension of every thing and being: the logic by which it posits itself as a self-existing thing is the very same self-causing causality that Spinoza calls God. God is therefore not a being-creator, but the force actively expressed through the existence of every thing.

Whereas this may sound theological to the modern ear, Spinoza's heretical position becomes clear in the first part of the *Ethica*, which is devoted to reasoning against the Christian conception of equivocal substance.[30] Based on the principle of reason, Spinoza derives the logical argument that there cannot be more than one substance. If there were several substances, as Descartes argues, their relationship would have to be determinate, meaning that the determined substances are not really substances at all, for they would not then be cause of themselves but of others. And if there is only one substance, there cannot be any difference in sense between creator and created. That is, the relation between God and Nature is *univocal*—God is said in the same sense of "himself" and his beings.[31] In fact, from Spinoza's perspective, God is not a unified being; it would be enough to say that God is vocal, that is, Voice itself, for it implies the same kind of necessary unity: "Whatever is, is in God, and nothing can be or be conceived without God" (IP15), which means that "God is the immanent, not the transitive, cause of all things" (IP18). And in turn, this means that God is another name for Nature, which inverts the hegemonic configuration of Christian theology and modern science: *Deus sive Natura*—God, or Nature, the two rendered metaphysically equivalent.

Like Descartes, Spinoza defines mind in terms of thought, or ideas, and body in terms of extension. Yet his explication of the mind-body relation in the second part of the *Ethica* inverts the Cartesian doctrine. Whereas Descartes argues that the act of thinking can verify our certain existence as independent minds, Spinoza argues that "the first thing which constitutes the actual being of a human mind is nothing but the idea of a singular thing which actually exists" (IIP11). This singular, actually existing thing is not the *cogito*: "The object of the idea constituting the human mind is the body" (IIP13). In other words, in Spinoza's view, Descartes's idea of his own thinking affirms nothing but the existence of that from which he believed his mind to be independent— his own body: I think, therefore my body exists. The mind and the body are two simultaneous expressions of the same reality. Contrary to Descartes, there is no "interaction" between mind and body, and consequently the "mind-body problem" that has riddled philosophy since Descartes does not occur in Spinoza's world-picture.

Spinoza's "orderly thinking" from the principle of reason leads him to refute widely held ideas of things that are considered "universals," such as the Christian (and Cartesian) freedom of the will. From Spinoza's perspective, universals are really "metaphysical beings," formed from thoughts about what the body encounters and, through the idea of these affections, turned into something without relation to the object. Such notions, he says, tend to be "confused in the highest degree. Those notions they call *Universal*, like Man, Horse, Dog and the like" all derive from the limited capacity of the body and the ideas of its affections:

> For the body has been affected most forcefully by what is common, since each singular has affected it. And the mind expresses this by the word *man*, and predicates it of infinitely many singulars. . . . But it should be noted that these notions are not formed by all in the same way, but vary from one to another, in accordance with what the body has more often been affected by, and what the mind imag-

ines or recollects more easily. . . . [E]ach will form universal images of things according to the disposition of his body. (IIP40S1)

In this sense, the disposition of the body or its affections constitutes the limit condition of reason. There is nothing about the structure of the human mind or its relation to the body that guarantees clear and distinct ideas. On the contrary, due to our inevitable dependence on the affects, the tendency of the mind would rather be toward the "mutilated and confused knowledge" that arises from encounters with singular things, from signs, opinion, and imagination. We are always affected by the world, and our mind is such that it invents causes of these affects under the principle of identity. Certainly, we are capable of reason and shared truths, but because we can never truly speak without affection—an idea as absurd to Spinoza as the void—we are always in some way "bound" and thus incapable of attaining pure universal knowledge. Spinoza offers us no guarantee for our knowledge of the world, as we cannot escape its forces.

The stubborn discrepancy between thinking and being, which in Galileo is made external to nature and in Descartes is made internal to the mind, becomes in Spinoza internal to Nature as such. In the fifth and final part of the *Ethica*, Spinoza defines human freedom and the ability to form clear and distinct ideas of the essence of things in relative rather than absolute terms. Freedom is not a given, as in Descartes's free will, but an always delimited attainment: "Each of us has—in part, at least, if not absolutely—the power to understand himself and his affects, and consequently, the power to bring it about that he is less acted on by them" (VP4S). Or as a subsequent principle puts it: "So long as we are not torn by affects contrary to our nature, we have the power of ordering and connecting the affections of the body according to the order of the intellect" (VP10). What we call reason is something we can train and improve, but we can never achieve certainty of whether we are not in some way "torn by affects contrary to our nature." Against the affects, reason alone is no guarantee, which is to say, light is the condition of both darkness and itself. The world is essentially complex and chaotic, and our powers, albeit considerable, are necessarily limited.

In turn, the analogy with truth sheds light on some political implications of Spinoza's logic. If truth is the standard both of itself and the
false, this means that "all ideas, insofar as they are related to God"—
that is, insofar as they are expressions of the same substance—"are
true" (IIP32). Contrary to Descartes's initial division of the world into
truth and falsity, Spinoza says that "there is nothing positive in ideas
on account of which they are called false" (IIP33). From the perspective of reason, this means that truth is fundamentally mediated by belief. If today's prevailing debates on realism and anti-realism turn on
the question of whether there are some statements about nature that
are really true, or only relatively true, or that there is no such thing
as truth, Spinoza argues that, as a matter of principle, everything that
someone believes to be true is true. This truth may be seen by others
as incorrect and may not be believed by them, but "he who has a true
idea, at the same time knows he has a true idea, and cannot doubt
the truth of the thing" (IIP43). Descartes's argument for the *cogito* is
in this sense reduced to an article of faith that may or may not convince others through the process of mediation. According to the same
(auto)logic, a believer of something believes it because he believes it.
Generally speaking, we argue, justify, and demonstrate not in order to
find the right belief; rather, belief is the initial spur in the turning that
becomes our arguments, justifications, and demonstrations. Thus, the
relationship between believing and knowing is analogous to the difference between the logics of initiation and turning: the voice speaks
truth even if the words fail to convince.

If reason is always mediated by belief, are we bound to a relativistic
slippery slope, or can we distinguish between true and false conceptions
in science? Since for Spinoza there is no positive, extrinsic criterion of
falsity, the distinction between truth and falsity becomes intrinsic to
scientific practice. That is, the determination of truth is always a matter of power. A sound and convincing argument certainly has power
in its mediation of interest, but not in a way that can be principally
distinguished from other forms of interest. Spinoza's work, in other
words, coheres with Stengers's picture of the political dimension of
the sciences. Interest for truth claims must be mediated and mobilized

among rivals and divergent actors whose own ideas, insofar as they are conditioned by their belief in the hypotheses, may possibly be as true as any other and whose fictionality is determined secondarily, in accordance with prevailing criteria for making history.

In general, Spinoza's theory of knowledge is predicated on the principle of reason, that thought always occurs on the condition of something different than itself and that it is itself therefore differentiated. *Semper sic*. Whereas Descartes's equivocal God guarantees the human mind its independence by which it can have universal knowledge of the world—in other words, that equivocity implies universality—we find in Spinoza the inverse conclusion, namely, that a univocal God implies a conditional knowledge upon which a *mathesis universalis*, an absolute, objective, and certain grasp of the cosmos, is impossible. The price to pay for reason is constitutive uncertainty.

THE MODERN METAPHYSICAL ORDER

In the passage from Galileo to Descartes on the one hand, and from Descartes to Spinoza on the other, modern metaphysics emerges out of two mutually exclusive modes of thinking. When the principle of identity trumps the principle of reason, we discover at the basis of metaphysics an ultimate identity, a being, a thing, a concept of God that by virtue of its identity implies a division from that which it is not. Thus, when the principle of identity is king, we find ourselves in an equivocal system, a separation of that which creates from that which is created. By contrast, when the principle of reason trumps the principle of identity, we find at the basis of metaphysics a connection, a relation, a self-cause that is not itself an identifiable self but whose essence can only show itself through subsequent identities, things, beings. Under the reign of the principle of reason we are in a univocal system, in which the creating and the created cannot be divided—they are but two expressions of the same substance, reducible only to a principle that connects. Like two inverse movements, Western metaphysics points in two different directions for all that follows. While Spinoza's way by the principle of reason will subsist in marginal philosophical discourses and immanent knowledge systems, the dominant thrust of modern metaphysics leans in the inverse direction, following a conception consonant with Christian metaphysics toward the messianic promise of *mathesis universalis*.

To be clear, the logical impossibility of a modern mathematical universe under an immanent metaphysics does not disqualify a quantitative universe as a useful and productive hypothesis. Rather, it simply means such a quantitative universe, like all hypotheses, will yield explanatory problems and ultimately have a limited reach.

Henceforth, the work of modern science becomes determined by this logical discrepancy between the world as it is unified by identity and the world as it gives itself. Developing on multiple, specialized fronts across disciplines and discourses, the modern scientific enterprise will be determined by the efforts to minimize the discrepancy within each framework until the framework itself becomes so incoherent that it requires reinvention in order to better account for the same stubborn discrepancy.

Historically, the Galilean-Newtonian model that began as a hypothetically empty universe for the purposes of calculation soon turned into a de facto empty universe. In 1788 the French mathematician Joseph-Louis Lagrange came up with a generalized expression of equivalence and equilibrium that, as Stengers puts it, took Galileo's invention out of the laboratory and "escaped" along with it into a purely mathematical realm. As with the case of statics, everything hinges on being able to render cause in terms of effect, to establish an identity through the mathematical marker of "=". Lagrange pushes this idea further: if the effect of a force can be determined, that force can in principle be replaced by any force with an equal effect, which allows for a whole new system of equivalences. As Stengers writes:

> Just as (real) equilibrium had been the instrument for defining the relationship between force and acceleration for Galileo, Lagrange used (fictional) equilibrium to define, for a given system characterized by a spatial configuration of a given collection of point masses, all the equivalent sets of forces, that is, all the sets determining the same acceleration for each component of the system. . . . The Lagrangian transformation is defined by the fact that it allows the

local equilibrium between (reversed) forces and acceleration to be expressed in terms of *independent spatial variables.* It amounts to representing a constrained system and the forces that determine its evolution as if it were a system of free points to which a set of fictional forces is applied.[32]

In other words, the given physical forces that Galileo had cleverly managed to turn into a hypothetical construct here finds its mathematical complement. And while few practicing physicists today pay much heed to Galileo, Lagrange equations have since become basic tools of not only theoretical physicists but economists and other modelers, because they enable a deceptive definition of mechanics as a thoroughly independent and neutral model, useful for generalizations across problems and disciplines. It enables a redefinition of motion in terms of successive states of equilibrium, in which each and every "moment" is fixed by the spatial configuration in which it occurs and indefinitely repeated. The statical model of equilibrium, which philosophically leads to questioning the cause of a certain state, turns under the regime of dynamics into mathematical equivalence. When the equilibrium of motions is understood abstractly and independently of forces, it offers the tantalizing possibility of disconnecting cause and effect of the statical moment of balance from any causal reference to a particular past or a particular future. As Stengers points out, Lagrange's model means that "cause and effect are reciprocally self-determining. Cause is not responsible for effect, it does not bring about the effect: its identity is derived solely from the relationship of conservation it shares with that effect."[33] A = A, indefinitely extended into a network of coordinates. The mathematical treatment of motion turns it into an endless series of successive instants, each defined statically by mobilizing the principle of identity through a fictionalized concept of equilibrium, effectively erasing what is always already deviating from this state, its given autological condition, by a hypostasized equation. Henceforth, as Stengers puts it, it becomes possible to forget the dynamic singularity in which any motion appears "to the benefit of some 'norm': the only truly intelligible phenomena are those that satisfy the conditions of the equation."[34]

From Galileo via Newton to Lagrange, then, the cosmos is reciprocally captured, purified, and emptied. Descartes's vortex has disappeared completely from physical view. Thus, in the history of the early consolidation of a new scientific culture in western Europe, it is perhaps Descartes, despite his brief period of immense influence, whose fate is most tragic. In his attempt to overturn the metaphysics of the Aristotelian order, Descartes showed ambitions that were too Aristotelian for the new science. On the one hand, his hydrostatical model of the universe proved impossible to quantify and thus failed to constitute a total mathematical framework for physics. On the other hand, after Newton's rise, and especially after Kant's reconfiguration of metaphysics, by the nineteenth century the natural philosophical project was no longer in need of its own metaphysical grounding. If Descartes's method initially served as a legitimatory philosophical narrative for bridging the fraught relationship between a new kind of science and an old church order, the new bifurcated order of modern scientific culture would eventually turn Descartes himself into the excluded middle.

In the new political constellation, in which natural philosophy and natural theology became mutually reinforcing, what was to be called science could effectively circumvent philosophical considerations and, on account of the Christian doctrine of revelation, inquire directly into nature. As Gaukroger describes it, "the idea that natural philosophy is a means of seeking evidence of God's activity in nature would become widespread in the 1680s and 1690s, particularly in England, and Newton for example would consider the stability of planetary orbits to be evidence of God's constant intervention." Natural constants became evidence for God's constancy. With increasing force, natural philosophy and revelation combined in the pursuit of what it saw as a shared truth—with theology as its initiatior:

> The kind of momentum that lay behind the legitimatory consolidation of the natural-philosophical enterprise from the seventeenth century onwards, a momentum that marked it out from every other scientific culture, was generated not by the intrinsic merits of its programme in celestial mechanics or matter theory but by a natural-theological imperative.[35]

Logically speaking, they were a perfect match, insofar as both pivot on the fundamental principle of identity. The God of natural-theological revelation is the identifiable creator—the being who produces Nature as universal. Natural philosophy, with Newton as its paradigmatic thinker, inquires into Nature as a creation of given, identifiable things. Thus, God gives us the universe, and the task of natural philosophers is to examine its parts, in ideal harmony.[36]

To be schematic, in the metaphysical configuration of equivocal universality, theology guarantees the ontology that natural philosophy can reconstitute epistemologically by way of logical, symmetrical division. The universe in its modern scientific sense is a symmetrical proposition that becomes axiomatic to scientific practice, because henceforth, modern scientists can substitute or complement their belief in God with the belief in mathematical universality—a belief which in turn legitimates the universality of the enterprise as such.

For Spinoza, on the other hand, who follows the logic of mediation through every aspect of his thought, the universe (in its Galilean-Newtonian or Cartesian form) is a fiction—or, as he would put it, a metaphysical being, a being of reason, which is confused for a real being. Yet, as he would have to admit, this confusion is not therefore simply an untruth. In fact, as metaphysical beings become taken for real beings, they are enacted into existence and thus in turn act into Nature. That is, like an undercurrent, they come to change the very conditions for reasoning.

In the case of the emergent modern world, this leads to a problem whose implications the *Ethica* is unable to account for. For what happens when the confusion between the two logical principles is not resolved but instead becomes generative of a new logical form? In the stubborn discrepancy between fictional universality and autological reality, which will preoccupy scientific practice for the next few centuries, something new is also mediated into existence. As a new kind of metaphysical being, it would come to exacerbate the tension in the metaphysical enterprise of seventeenth-century natural philosophy and make the legitimatory circulation between science and religion much more complicated. The inklings of change were found in Spinoza's own time, but not until the late nineteenth century does it appear with such force that both physics and philosophy, amid a sea change of disciplinary knowledges, are forced to reinvent themselves in order to handle the consequences.

Thus, from the early logical foundations that invented the universe as we know it, we need to jump to the next metaphysical revolution in the history of modern cosmology.

3

BEYOND THE OPTICAL LIMIT: Neptune, hidden giant to the naked human eye.
Copyright NASA, Creative Commons.

PROBABILITY AND PROLIFERATION
The Invention of the Particle

To philosophize means to reverse the normal direction of the
workings of thought.

—*Henri Bergson*

METALOGICAL GROWTH

In 1787, an entirely new planet was discovered by telescope and was an
instant cosmological sensation. Since ancient times, Saturn had been
visible to the naked eye and known as the most distant planet in the
sky. The new planet, named Uranus, was enormous, many times larger
than Earth and clearly a major hidden force in our cosmos.

Decades later, based on observations and measurements of its
gravitational path using Newton's theoretical framework, astronomers
were able to predict that some other large object was affecting Uranus.
On the basis of these calculations, another and even larger planet, twice
the size of Uranus and six times the size of Earth, was finally observed
in 1846. Dubbed Neptune, its discovery expanded the known universe
even further, allowing for grand conjectures about what might be
found farther out in the cosmos. But a certain limit was being reached.

After the first major leap of microscopy and telescopy in the 1600s,
the single-lens contraption that had enabled large-scale magnification
reached its apotheosis in the latter half of the nineteenth century. In
1872, lens maker Carl Zeiss hired Ernst Abbe, a professor at the University of Jena to improve a compound lens apparatus. Aided by an
advanced mathematical understanding of lenscraft based on the

prevailing wave theory of light, Abbe was able to calculate the theoretical limit condition of microscopy, which could be applied to telescopy as well. According to his formulation, the physical limit of microscopic resolution amounts to half the wavelength of light, which for light in the visible spectrum corresponds to a maximum object resolution of around 0.3 micrometers. Under a theoretically perfect lens, then, any two lines closer together than this limit will appear as one, and any object smaller than it will be invisible or indistinguishable. In the biological realm, this limit lies somewhere between the smallest bacteria and the largest types of virus. In effect, Abbe invented the theory of aperture that is still the basis of photography today—and Zeiss's lenses are still industry-leading. In the following decade and a half, Abbe and Zeiss approximated this theoretical limit experimentally by producing a series of advanced compound microscopes. In 1889, Abbe made a lens with the highest numerical aperture ever manufactured. Capping out at roughly one hundred times the magnification of the microscopes in the 1600s, the optical light microscope had for all practical purposes advanced to a level of resolution it has never since been able to surpass, at once maximizing the continuity of light and reaching its limit. Faced with this profound constraint, where was science to go?

What was at stake for physics in the early twentieth century was inventing the means to proliferate past its own physical limitations—to extend its reach beyond its own metaphysics. In the nineteenth century, the idea of other solar systems or galaxies outside our own was still an unthinkable leap. Despite the discovery of more distant planets, the universe was essentially the same as before, unified and whole within a set of observable laws that proved remarkably precise. When such a giant leap to new galaxies eventually occurred (which the next chapter is about), it was not merely due to advancement in telescopic technologies. In order to understand the cosmology of the twentieth century, we must first see how it was born out of a metaphysical revolution in late-nineteenth-century physics and mathematics that yielded a new form of reasoning, invented a new fundamental physical reality, and developed a complicated set of logics that would eventually transform our idea of the universe.

The first claim to have discovered an elemental particle (the electron) occurs in this context at the very end of the nineteenth century. To render an atomic level of still highly hypothetical matter constituents visible to the human eye would require at least one hundred times magnification beyond Abbe's limit. Thus, the discovery of atoms, which transformed twentieth-century physics, did not occur determinately, as a particle rendered visible under a microscopic lens, like Neptune, but rather probabilistically. That is, an array of uncoordinated experiments under highly specialized and qualified conditions appeared to mutually indicate a variety of mathematically stable phenomena whose implicit demand for a unifying explanation warranted invocation of a particle concept as an operational hypothesis for further research. If that sounds complicated for experimental proof, it's only the simplified half of it. For what emerges as a physical particle, as we will see, is a distinctly two-faced construction.

This metaphysical revolution was certainly not of Spinoza's making, and Spinoza might have conceived it as contrary to his rational system, because it emerges logically, as we will see, beyond the principle of reason. Nevertheless, we can find inklings of its emergence foreshadowed in his own work. In the first part of the *Ethica*, Spinoza distinguishes between what he calls *Natura naturans* and *Natura naturata*, between "naturing nature" and "natured nature," loosely translated, active versus passive nature. For Spinoza this is only a conceptual separation, as we are, in any given moment, simultaneously active when we affect others and passive when we ourselves are affected. By active nature, Spinoza understands substance itself, that is, "God, insofar as he is considered as a free cause"—this is Nature as Essence, its striving to persist in its being. By passive nature, Spinoza understands nature as affected modes, the effects of such striving activity, that is, Nature as existence (IP29S). These are in Spinoza's system two simultaneous and coexisting logical dimensions of the same reality—to remove one is to consider the world without causality, and to remove the other is to consider a world without actual effects. Thus, hypothetically, if we only consider Nature insofar as it is *naturata,* passive nature, we encounter a realm without the axiomatic, autological principle of reason—a realm of pure contingency and chance.

In 1657, Descartes's follower Christiaan Huygens, encouraged by the French philosopher Blaise Pascal, published what today stands as the first textbook on probability reasoning, *Calculating in Games of Chance*. It's as good an origin marker as any, if only because it appears to be the first of historically preserved texts to make it to the printer. As Ian Hacking points out, the emergence of probability as a historical phenomenon has no clear, singular origin but rather seems to spring up independently in textual records around the same time in the Netherlands, in Paris, and in England. Besides Huygens, Pascal worked on an early conception of probability in a different vein, as did Leibniz independently. And in London, statistical records were compiled from which new mathematical patterns were inferred—all in the course of the 1660s. From this general emergence to the eighteenth-century formal problematization of induction (David Hume) and statistical reasoning (Thomas Bayes), there appears no direct causal path, which makes the historical emergence of probability itself, appropriately perhaps, akin to a statistical phenomenon.[1] As a contemporary of Huygens, Spinoza participated in study circles that were discussing, apart from natural-philosophical subjects like optics and physics, the reasoning and mathematization of chance. In correspondence with Huygens in the mid-1660s, at the same time he was at work on the *Ethica*, Spinoza dabbled with some of the problems in Huygens's book and wrote a short treatise on the subject himself, *Calculation of Chances*. Much like Huygens, Spinoza approaches the calculation problem in a statical way. First an equivalence is established (Huygens calls it "equivalent gambles")—in every coin toss, for example, there is an equal chance of yielding heads or tails—from which a series of generalized equal chances can be inferred and thus calculated. But theirs were only two of many divergent approaches to understanding the new realm of chance.

From its inception, Hacking writes, probability has been two-faced. "On the one side it is statistical, concerning itself with stochastic laws of chance processes. On the other side it is epistemological, dedicated to assessing reasonable degrees of belief in propositions quite devoid of statistical background."[2] He provides a brief overview of the history of this problem, which has preoccupied students of probability reasoning for centuries. Various attempts have been made to either distinguish the two aspects by name—for instance, chance on the one hand, credibility on the other—or, going the opposite way, to claim that one aspect of probability is a subset of the other. As Hacking relates, none of these approaches succeeded, and clearly two families of reasoning remain analytically distinct, even as their origin is shared.

> Philosophers seem singularly unable to put asunder the aleatory and the epistemological side of probability. This suggests that we are in the grip of darker powers than are admitted into the positivist ontology. Something about the concept of probability precludes the separation which, Carnap thought, was essential to further progress. What?[3]

Hacking's study is not devoted to answering his own question, though the suggestion that "we are in the grip of darker powers" is perhaps more instructive than intended. After all, as an analytic branch of philosophy, modern epistemology, like modern physics, operates within a logical space predicated on the circumscription of the autological. Without implicit reference to how logic emerges from a positive given reality, I would argue, the difference between the two dimensions of probability reasoning becomes difficult to identify. Let us briefly consider these two faces of probability.

On the epistemological side, the two dominant schools of thought are both concerned with the probability conferred on a hypothesis by some evidence, conceived either as a certain relation between two propositions or as a matter of personal judgment subject to rules of internal coherence. How credible is a person's hypothesis? How probable is its prediction? As Hacking puts it, "no matter whether the logical or personal theory be accepted, both are plainly epistemological, concerned with the credibility of propositions in the light of judgement or evidence."[4] Epistemic probability, in other words, is concerned with the general scientific problem concerning the discrepancy between the autological given and the conceptions of identity. Its objective is to find a way to measure a degree of correspondence between a concept—an invented origin—and its experimental occurrence, that is, between a concept and its circumscribed reality. In this sense, an autological dimension is involved insofar as there is causal correspondence. Moreover, the autological is involved insofar as the analysis of the probability relation—be it extrinsic, between two propositions, or intrinsic, in terms of personal judgment—explicitly involves belief. Whether a hypothesis is considered believable in general or believed by the subject involved, we are still nominally tied to the autological dimension from which modern science attempts to remove itself. Epistemological probability, in other words, is based on a determinate relation of causality.

On the aleatory side, the dominant theories focus either on the problem of randomness in infinite and finite sequences—in other words, on effects—or on the causes of frequency phenomena, conceived as the propensity or tendency for a test to yield one of several possible outcomes. "Clearly," Hacking writes, "none of this work is epistemological in nature. . . . [T]he stable long run frequency found on repeated trials is an objective fact of nature independent of anyone's knowledge of it, or evidence for it."[5] Insofar as the work is mathematical in a general sense, aleatory probability is also clearly governed by the principle of identity. However, if it can be considered non-epistemological, as Hacking claims, it is not because it is "objective" or "physical" but rather because it does not involve a determinate relation between a concept and its reality. In statistical reasoning, we do not invent an origin whose reality we circumscribe; rather, inversely, we invent a reality within which we can circumscribe an origin. As students of statistics know, correlation is not causation. And if no causality is involved, as per the principle of reason, neither is the autological. A chain of disconnected numbers shows their own pattern.

In common practice, it is convenient to distinguish aleatory probability as involving large sets of numbers and referring to macro-level events. But the distinction between the two faces of probability is not quantitative in nature, nor is it strictly a matter of scale. There is no identifiable threshold or limit where epistemic probability ends and aleatory probability begins—nor is there, as the many historically unsuccessful attempts indicate, a way to make the two dimensions correspond. Rather, their distinction could be considered logical in the way I define it here, determined by their relation to the autological. Thus, on one side we find the autological is implicit, turned around and circumscribed. On the other side, the autological is excluded altogether. Because it is noncausal, it goes beyond the autological, beyond the principle of reason as its pivot. This kind of reasoning becomes what I call *metalogical*. Whether we think of *meta-* in denotative terms of "beyond," "after," or a positional change, the metalogical signifies a kind of reasoning able to turn away from its autological ground and operate on its own terms.

Metalogic is acausal, that is, without cause in the generalized sense. As contemporary physicists might put it, it is "nonlocal," not linked to a specifiable, localizable causal trajectory. But metalogic is certainly not without effect. Hacking's *The Taming of Chance* is a general study of the spectacular rise of probability in the nineteenth century—and in particular its statistical side. Recounting what he calls an "avalanche of numbers" published after the Napoleonic Wars, from birth and death rates to insurance premiums and civil status, Hacking explicitly links the rise of statistical reason to the French philosopher Michel Foucault's notion of biopolitics, involving the extensive mapping, control, and manipulation of populations within bounded nation-states, based on what came to be regarded as autonomous statistical laws, in turn a precursor to today's algorithmically based social media. These are laws that would deeply affect the modern world.

> Statistical in nature, these laws were nonetheless inexorable; they could even be self-regulating. People are normal if they conform to the central tendency of such laws, while those at the extremes are pathological. Few of us fancy being pathological, so "most of us" try to make ourselves normal, which in turn affects what is normal.[6]

In this sense, quantitative extension thus implies qualitative intension—a "feedback effect," as Hacking calls it. This feedback effect, crucial to understanding many social processes, is also crucial to the conceptual distinction of metalogic from statistics as such. For it is first and foremost through the aggregated feedback effect enabled by statistics that something distinct from conventional logic occurs. Beyond an initiation and its folding, beyond the ceaseless ping-pong of discrepancies between reality and concept, something emerges along with this new logical order that can be differentiated as a vector for itself. The metalogical signifies increased mobilization, acceleration, multiplication, viral logic.[7] Much begets more.

As Hacking points out, effects of statistical reason clearly occur in a social sense, as his example of the self-regulation of normalcy indicates. However, he then makes an important differentiation: "Atoms have no such inclinations. The human sciences display a feedback effect not to be found in physics."[8] Malleable humans—but permanent atoms? In the thematic context of biopolitics, the distinction between physical and human nature appears commonsensical, but nevertheless it is questionable. Historically, it mirrors the identity of human as differentiated from nature and thus the modern configuration of the universe itself: nature, as differentiated from culture, is the realm independent of human knowledge, and which in turn sustains a notion of the independent human knower. Classical humanism lurks in the background.

In this chapter I will argue there is a fundamental feedback effect in physics that goes largely unacknowledged in modern thought. Perhaps this is partly because the feedback effect in physics runs inversely to its manifestation in the human sciences. It crucially relies on the concept of particularity—the idea of individual or independent identity. In his charting of the social conditions for statistical reasoning in the nineteenth century, Hacking makes a distinction he describes as "gross but convenient" to illustrate a general point:

> Statistical laws were found in social data in the West, where libertarian, individualistic and atomistic conceptions of the person and the state were rampant. This did not happen in the East, where collectivist and holistic attitudes were more prevalent. Thus the transformations that I describe are to be understood only within a larger context of what an individual is, and of what a society is.[9]

Why would statistical laws occur more readily in connection with individualism? Because without defined individuals—that is, discontinuous, identifiable units—metalogic is impossible. The reliance on particularity signifies the debt to the principle of identity. When the identity of cause and effect is relinquished, the principle of identity remains as though it has been overturned, now existing only by implication, as a negative presupposition. Statistical reasoning presupposes individuals.[10] This is one reason why, for example, opinion polling functions so readily in liberal democracies, in which the individual political subject, identified as an independent voter, is already invented and, through the subsequent rise of statistical reason, reinforced. In the case of physics, the logic runs the other way, since what was required to implement statistical methods and to take them beyond physical limits and to expand the universe as such was precisely the invention of distinct individual units.[11]

MAXWELL'S CONTINUITY

The story of the metalogical leap in physics has, as perhaps befits it, a rather proliferating plot. If this history can be narrowed down to a few pivotal characters, it is the Scottish physicist James Clerk Maxwell's work that comes to play a crucial double role—much as Einstein's work would provide a decisive double juncture four decades later. From Maxwell through Einstein and a few key figures in between, we find the contours of the paradoxical features that will bedevil physics for the following century.

In historical terms, the development of probabilistic reasoning is perhaps better characterized as a revolution in application rather than in science. As its documented historical emergence in the mid-seventeenth century indicates, probability sneaks up on the modern world so slowly, and through such a multitude of separate events, that it can be considered revolutionary only in an extended, century-long sense. The increasing regularity within the "avalanche of numbers" corresponds to an exponential vector. In concert with several biopolitical measures, population figures in Europe take on their characteristic runaway effect. Within less than three generations, between 1850 and 1910, the great metropolitan centers of Europe roughly triple in population. Joint-stock companies rising to dominate national economies occur alongside unprecedented mobility of labor for increasingly mechanized forms of work. Greater populations require greater metalogical measures, which improve conditions for increasing populations, in ever greater feedback correlations between modern science and politics.

However, partly due to its deep mechanist inheritance, physics takes up statistics late. As I explained in the previous chapter, Newton's framework of dynamics was predicated on a foundational circumscription of what natural philosophers of the seventeenth century had defined as force—the presence of an autological given. Through kinematics, the problem of gravity was transformed into a proportional constant between two bodies, G. While "cosmology" at the time was restricted to observations and basic calculations of events in the sky, the development of physical science in the eighteenth and nineteenth centuries was analogous to an experimentally layered refilling of the empty Newtonian spatiotemporal container called the universe. Physics, which now takes shape as a recognizable, institutionalized discipline, becomes dominated by questions of the so-called imponderables, which Newtonian dynamics could only outline, from light, heat, and gas, to sound, energy, and electricity—all "the stuff of the world"—which brings forth new problems according to the increasing discrepancy between the universal framework of dynamics and actual observable conditions.

The most problematic field was thermodynamics. Developed from the physics of heat processes, thermodynamics establishes an equivalence between heat and work (using Lagrange's mathematical invention) through the concept of energy. In 1865, the German physicist Rudolf Clausius formulated the two laws of thermodynamics. The first law states that "the energy of the universe is constant." Thus, energy takes on a metaphysical meaning in direct analogy to the idea of substance—a name for that which does not itself change, but enables any process in the universe. In a conceptual distinction that harks back to Aristotelian metaphysics, energy is subdivided into potential and actual (later called kinetic) energy. As this total of actual and potential, energy is an expression of autologically mediating presence. Thus, in a radical inversion of the kinematic law for a falling body in a void, the first law of thermodynamics constitutes the world as a plenum, a full universe with a constancy of energy.

For the second law of thermodynamics, Clausius proposed the Greek word for transformation, *entropy*, to describe the ostensibly irreversible, directional character of heat flowing from hot bodies to colder ones: "the entropy of the universe tends to a maximum." In this way, entropy also expresses metaphysical constancy, because it is defined as a tendency toward a limit condition. In both laws, as Stengers observes, a cosmological limitation is implied by the condition of the "universe," which does not indicate the absolute character of the laws but rather the opposite: the universe is "the only 'system' that, by definition, does not undergo exchange with an environment. That is why it is the only case that allows energy and entropy to be subject to statements of similar scope."[12] The extent to which energy is conserved or degraded, just like in the case of Galileo's motion, always depends on actual physical circumstances that are highly variable—and so the empty universe provides Clausius with the exceptional and fictional reference frame necessary to state his laws as absolute.

Taken together, these two laws of thermodynamics describe what Clausius, Maxwell, and other physicists at the time took to be the fundamental features of energy, which is itself turned into a fundamental concept of conservation and dissipation. Everything is in equilibrium, on condition of disequilibrium—everything begins with balance, on condition that it tilts. In this sense, thermodynamics is a theorem of statics (as I described in the previous chapter) that replays the constitutive difference in dynamics between a kinematical void and a statical plenum. Rather than explaining motion in the absence of force, as Galileo did, statics and thermodynamics describe force without motion, as tendency.

Moreover, the two laws of thermodynamics express the extensive relation between part and whole. The second law explicates the tendency of any part toward the whole that is constituted by the first law. That is, entropy is the qualitative extension of energy, from the dissipation of its parts to the conservation of the whole.[13] For Maxwell, the physical promise of the concept of energy and its formulation through the laws of thermodynamics lay in its potential to replace the traditional but mathematically imprecise notion of force altogether. Several key works of the 1870s espoused the view that Newton's "abstract" dynamics could successfully be derived from thermodynamics as a new basis for physical explanation. In turn, energy was established with the same fundamental reality status as matter itself. But how to reconcile matter with energy?

Much of the development of nineteenth-century physics revolved around this metaphysical problem. The atomistic conception of matter, as fundamentally constituted by indivisible particles, had its adherents, encouraged by developments in chemistry that theorized individual chemical elements as atomic. But chemical atoms (elements) were an expression of the limit of chemical analysis, the point at which composite bodies break down, analogous to where two lines under a microscope become one. By contrast, a coherent physical conception of the atom would require something so fundamental as to be invisible to experiment. The problem was not that matter could be regarded as a composition of molecules for hypothetical use, but rather how such molecules could be considered truly fundamental. Physical explanations moved toward an autological understanding of the problem: if the Newtonian universe indeed consisted of bodies, or planets, in mechanistic motion, something was needed to mediate them. Irrespective of the molecular and quantitative models used to explain energy as a physical phenomenon, thermodynamics appeared to signify an absolute continuity of energy in the universe, both in terms of its totality (the first law) and its flux (the second law). Foregrounded in Maxwell's work, a set of analogous key concepts reinforces this idea of a positive continuity.

Mathematically, modern physics since Newton was significantly shaped by the invention of differential calculus. Used to formulate the basis of rational mechanics, a derivative function expresses a fundamental differential equation between variables that, if unlimited by other factors, can be extended indefinitely. Accordingly, the concepts of mathematical physics tend to express continuous functions. This was particularly the case with Maxwell's electromagnetic field, his proposal for the mathematical unification of the physical phenomena of electricity and magnetism. Based on Michael Faraday's proposition of lines of force, the general field theory allowed forces between individual bodies to be mediated by the continuous propagation of energy. In this sense, Maxwell noted, the continuous field theory proved a direct correlate to the prevailing wave theory of light, first formulated in the mid-seventeenth century by Christiaan Huygens, developed from the metaphysics of Descartes.

Thus, theoretical and experimental lines of inquiry, the coherence of mathematical equations, physical instruments, and metaphysical axioms, all led toward explanatory unification. In order to make the theory of light cohere with mechanics, electromagnetism, and the thermodynamic conception of energy, physics needed an autological medium to carry energy, light, and electromagnetic waves. It became known as the *ether*—a physical concept whose many varying interpretations belie its prominence in nineteenth-century physics. In 1878, Maxwell authored the entry on ether for the ninth edition of the *Encyclopedia Britannica*, arguing its case: "The evidence for the existence of the luminiferous ether has accumulated as additional phenomena of light and other radiations have been discovered; and the properties of this medium, as deduced from other phenomena of light, have been found to be precisely those required to explain electromagnetic phenomena." Maxwell further explains that in order to transmit energy, the ether would need to possess "elasticity similar to that of a solid body, and also have a finite density." He cites conceptual problems over whether the ultimate constitution of the ether is molecular—that is, contiguous elements through which undulatory light is capable of passing—or continuous in itself. Nevertheless, his conclusion is emphatic:

> Whatever difficulties we may have in forming a consistent idea of the constitution of the ether, there can be no doubt that the interplanetary and interstellar spaces are not empty, but are occupied by a material substance or body, which is certainly the largest, and probably the most uniform body of which we have any knowledge.[14]

To the late-nineteenth-century British physicist, in other words, the existence of the ether was indubitable—a necessary presupposition—for without it, the prevailing concepts of physics would cease to make coherent, unified sense within the framework of mechanics.

Over the next two decades, the discourse of physics would become rife with proliferating mechanical, thermodynamic, gaseous, molecular, particulate, and fluid models to explain this necessary autological mediation. Among the theories Maxwell worked on was a version of Descartes's idea of the vortex. In fact, one of the most promising and influential concepts in the 1890s for uniting the demand for particular discontinuity with a universal ethereal plenum becomes known as the vortex atom, advocated most prominently by William Thomson (later Lord Kelvin). Thomson's model, developed and synthesized with the latest experimental phenomena by Joseph Larmor, conceives of matter itself as vortex rings in a primordial fluid medium. Consequently, as Larmor put it, "matter may be and likely is a structure in the ether, but certainly ether is not a structure made of matter."[15] The ether, in other words, was conceived as ontologically primary, a pure continuum within which particulate matter occurs.

Perhaps the most eloquent expression of this general tendency in physics was captured by the French philosopher Henri Bergson in his *Matter and Memory*, first published in 1896. He summarizes Faraday's conception of centers of force and Thomson's vortex rings as two different explanatory attempts that coincide in a metaphysical vision of universal continuity:

> In truth, vortices and lines of force are never, to the mind of the physicist, more than convenient figures of illustrating his calculations. But philosophy is bound to ask why these symbols are more convenient than others, and why they permit of further advance. Could we, working with them, get back to experience, if the notions to which they correspond did not at least point out the direction we may seek for a representation of the real? Now the direction which they indicate is obvious; they show us, pervading concrete extensity, modifications, perturbations, changes of tension or of energy, and nothing else.[16]

For Bergson, the contemporary trend in physics was highly encouraging, because concepts such as the ether signified the acceptance in physics of the philosophical distinction between the continuous movement of reality and the artificial divisions of the mind, such as the popular but as yet unfounded physical theory of atoms. As Bergson puts it,

> We shall never explain by means of particles, whatever these may be, the simple properties of matter: at most we can thus follow out into corpuscles as artificial as the corpus—the body itself—the actions and reactions of this body with regard to all others. . . . But

the materiality of the atom dissolves more and more under the eyes of the physicist. We have no reason, for instance, for representing the atom to ourselves as a solid, rather than as a liquid or gaseous.[17]

He further cites two experiments conducted by Maxwell that effectively counter any conception of particles existing in a void. Thus, he concludes that modern science ends up with a concept like the ether to account for the autological mediation of its individualized particle concepts.

We see force more and more materialized, the atom more and more idealized, the two terms converging towards a common limit and the universe thus recovering its continuity. We may still speak of atoms; the atom may even retain its individuality for our mind which isolates it; but the solidity and the inertia of the atom dissolve either into movements or into lines of force whose reciprocal solidarity brings back to us universal continuity.[18]

In this sense, the nineteenth century appeared to close on a dizzying note of advancement toward the limits of physical understanding. Prominent physicists began talking publicly about the end of physics, the near-completion of physical problems. And Bergson rose to the occasion with a remarkable and, above all, hopeful proposition: that this metaphysical limit was the condition under which philosophy and science could coexist in a new conceptual alignment.

BERGSON'S HARMONY

At the turn of the twentieth century, as science and philosophy had grown ever more distant, reconciling metaphysics and science was already a monumental task. As Bergson remarked in his 1903 essay, "An Introduction to Metaphysics":

The masters of modern philosophy have been men who had assimilated all the material of the science of their time. And the partial eclipse of metaphysics since the last half century has been caused more than anything else by the extraordinary difficulty the philosopher experiences today in making contact with a science already much too scattered.[19]

Bergson's *Matter and Memory* from 1896 is one of the last major comprehensive attempts in the Western canon at making viable contact between physics and philosophy. His work combined a critical renewal of modern philosophy's premise with conclusions that proved remarkably convergent with the most advanced physics of his day. However, Bergson was not a believer in *mathesis universalis*. Despite its attempt to reconcile the foundations of psychology, biology, physics and philosophy in one and the same work, *Matter and Memory* shows no pretension to theoretical unification. Instead, Bergson calls for a dualism—but an uneasy yet mutually constitutive dualism that shares overtones with Spinoza. In this sense, Bergson writes both with and against the traditional current of modern philosophy, which in turn has made his thinking eminently difficult to categorize.

Matter and Memory begins with a strictly conceptual fiction called "pure perception"—an inversion of both Descartes's *cogito* and Kant's concept of pure reason. Bergson invites us to think of our experience of the world as an unlimited fullness: "In pure perception we are actually placed outside ourselves, we touch the reality of the object in an immediate intuition."[20] This world of pure perception consists only of what he defines as images. With this conceptual invention, Bergson intends to navigate between, on the one hand, the "representation" of idealist thought, and on the other hand, the "thing" of realist thought. In a manner analogous to how Descartes describes the world from the point of view of the *cogito*, Bergson says that when we look at the world from the perspective of pure perception, "all seems to take place as if, in this aggregate of images which I call the universe, nothing really new could happen except through the medium of certain particular images, the type of which is furnished me by my body."[21]

Of course there is no such thing as pure perception, which Bergson acknowledges, since perception is always and everywhere mixed to some degree by bodily affection. But as an operational premise it enables him to circumscribe the world in terms of a plenum rather than a void: matter consists of an aggregate of images, whereas perception of matter—that is, the interface of mind upon matter—consists of the same images as limited by that one particular, privileged image called my body. Bergson's premise is distinctly Spinozist, in that one and the same univocal reality is expressed in terms of two attributes, or in Bergson's language, two systems of images:

> Here is a system of images which I term my perception of the universe, and which may be entirely altered by a very slight change in a certain privileged image, my body. This image occupies the center; by it all the others are conditioned; at each of its movements everything changes, as though by a turn of a kaleidoscope. Here, on the other hand, are the same images, but referred each one to itself; influencing each other, no doubt, but in such a manner that the effect is always in proportion to the cause: this is what I term the universe. The question is: how can these two systems co-exist, and why are the same images relatively invariable in the universe, and infinitely variable in perception?[22]

Conceiving his problem this way, Bergson attempts to circumvent two related pitfalls of the universe in modern philosophy. The first concerns ontological bifurcation, and the second revolves around epistemological predominance. In both cases, Bergson's orientation aligns with Spinoza's. As I discussed in relation to Descartes's *Principia*, a prototypical first move of modern philosophy is to create an ontological distinction of symmetry—between interiority and exteriority, between subject and object, or in Descartes, between mind and body. This difference in kind is predicated on the principle of identity, by which a discernible difference becomes a marker of separate identities—mind on one side, body on the other. And once this difference in kind is established, it leads to, as Bergson notes, a problem whose terms are strictly insoluble by themselves. Either we are left with a "mysterious correspondence" between the two symmetrical systems—for which a deus ex machina, such as a miracle or preestablished harmony, becomes necessary—or we disavow transcendental solutions altogether and instead forge from the relational question a causal explanation: how the outer physical world causes the inner mental world of the subject, or vice versa. As Bergson succinctly puts it, "subjective idealism consists in deriving the first system from the second, materialistic realism in deriving the second from the first."[23]

Instead, Bergson simply proposes that inside and outside, or subject and object, are functions of image relations—not the other way around. And because the difference between the two systems emerges in the limitation of the body, their coexistence is not symmetrical but rather asymmetrical. It corresponds, he says, to the distinction between part and whole. In this sense, Bergson's philosophy can be read as a double analysis of how the body moves from its own image system to the universe by extension, while simultaneously changing in itself by intension.[24]

Upon close reading, the principal differences between Bergson's and Spinoza's metaphysics consist in terminology. Metaphysically, Bergson's two systems cohere with Spinoza's difference in attributes between a body and the idea of the body. That is, what Spinoza in proper rationalist form calls "thought" is for the more empirically oriented Bergson "perception." For Bergson, "to perceive means to immobilize" in much the same way that for Spinoza thought without relation to its object—that is, the idea of the idea of affection—is a mental construct, a hypostasis of mind. Their conceptual convergence becomes clearer through a metaphor that Bergson draws from optics:

> When a ray of light passes from one medium into another, it usually traverses it with a change of direction. But the respective densities of the two media may be such that, for a given angle of incidence, refraction is no longer possible. Then we have total reflection. The luminous point gives rise to a virtual image which symbolizes, so to speak, the fact that the luminous rays cannot pursue their way. Perception is just a phenomenon of the same kind. . . . This is as much as to say that there is for images merely a difference of degree, and not of kind, between being and being consciously perceived.[25]

Yet this reflection is, in Bergson's continuation of the metaphor, a kind of mirage, since the virtual image conceals the fact that rays still travel through it. Thus the effect is, from the perspective of the behavior of light rays, "that the real action passes through, and the virtual action remains."[26] In fact, the image trope here falls short, because the body, as the privileged image of both systems, does not "merely reflect action received from without; it struggles, and thus absorbs some part of this action." This absorptive capacity, Bergson says, stands for the source of affection. Thus, "while perception measures the reflecting power of the body, affection measures its power to absorb."[27] In mediating the body, in other words, light effectively doubles in its movement: in reflection, it is moved "under" or turned on itself and at the same time continues through itself. For Spinoza, idea is the thought of an (autological) affect and also, in a doubling, the thought of this thought, hypostasized, as though by a reflected, virtual image. This doubling corresponds logically to the two image systems that Bergson differentiates: "The first system alone is given to present experience; but we believe in the second, if only because we affirm the continuity of the past, present, and future."[28]

Thus, the two systems constitute a double movement, their loops feeding simultaneously forward and backward. By privileging one system, we get caught in a paradoxical reliance on the other, for which the philosophical divide between realism or materialism and idealism is an exemplary expression. If we assert that a logical movement "really" or "naturally" runs from the autological given to the mind's identity constructs, we are soon forced to acknowledge that we can only make this claim on account of the inverse movement, using the mind's constructs to make sense of the given reality. Asserting the primacy of either system is precisely to flatten the double movement of mind and body, an evisceration that becomes a double error.

The rivaling doctrines that attempt to explain one movement in terms of the other do not merely begin with a false ontological bifurcation, Bergson argues. Together, they also mutually constitute a false epistemological image—the second pitfall of modern philosophy. With regard to the problem of so-called subjective knowledge, he writes:

> The whole discussion turns upon the importance to be attributed to this knowledge as compared with scientific knowledge. The one doctrine (realism) starts from the order required by science, and sees in perception only a confused and provisional science. The other (idealism) puts perception in the first place, erects it into an absolute, and then holds science to be a symbolic expression of the real. But for both parties, to perceive means above all to know.[29]

Insofar as *Matter and Memory* attempts to reconfigure the relationship between philosophy and science, metaphysics and physics, it tries above all to dispute the traditional postulate of Western thought that knowledge is the telos of perception. The enduring myth of science is that it is founded on the idea of knowing truth for itself, whether truth is understood in realist or idealist terms. For Bergson, however, as for Spinoza, the mind is not geared toward knowledge or truth for its own sake but rather toward action.

Bergson defines action as "our faculty of effecting changes in things," and he links it directly to the indetermination that is "characteristic of life." In fact, he says, living matter, or active perception, is defined by its ability to defy determinate expectation, to evade conceptual capture—and this trait constitutes its essential freedom.

> Perception arises from the same cause which has brought into being the chain of nervous elements, with the organs which sustain them and with life in general. It expresses and measures the power of action in the living being, the indetermination of the movement or of the action which will follow the receipt of the stimulus.[30]

The emphasis on life becomes a stronger motif in Bergson's later writings, most notably *Creative Evolution*, where the idea of action is developed into an account of vital tendency (élan vital). For Bergson, life is precisely what mechanistic thought misses—and in his careful distinction of action from affection, he appears to be distinguishing his own philosophy from Spinoza's. The reciprocal dependence of conscious perception and cerebral movement—that is, the correspondence between mind and body—is for Bergson "simply due to the fact that both are functions of a third, which is the indetermination of the will."[31]

At first glance, nothing could be further from Spinoza's trenchant claim in the *Ethica* that people's common "opinion" of having free will "consists only in this, that they are conscious of their actions and ignorant of the causes by which they are determined. This, then, is their idea of freedom—that they do not know any cause of their actions" (IIP35S). Certainly, in the Western canon, Spinoza is often considered the ultimate expression of seventeenth-century rationalist determinism—a sentiment radically alien to Bergson's indeterminist philosophy of life. Nevertheless, this common interpretation is based on a misunderstanding of how Spinoza and Bergson come to solve the problematic relation between the two systems, mind and body. As we have seen, both thinkers share the univocal premise that the coexistent systems are expressive of the same reality—that mind and body are the same images expressed in irreducibly and qualitatively different modes. However, for Spinoza this coexistence finds its ultimate alignment in God (or Nature) as an apotheotic principle of reason. His perspective develops from the metaphysical whole in order to explain how the individual body and mind participate in it. Bergson, on the other hand, approaches the problem from the inverse perspective—from the individually perceiving mind toward the greater whole.

Thus it should not be surprising that their divergent stances yield different conclusions. Spinoza's determinism is strictly a function of his perspective, which sees in the autological striving of each individual body an expression of the logical principle by which the world is immanently connected. Considered exclusively from the point of view of this connecting principle, nothing could be indeterminate or unconnected by it, since the principle by itself expresses the whole. But this is very different from claiming that our actions are determined by an identifiable being—recall that God or Nature in Spinoza is not a function of the principle of identity—or that individual minds could ever grasp the order of this determination, since this is precisely what he precludes. If we were to consider the world only from within the ever-changing order of phenomena, *naturata* (passive nature), the result is not ironclad scientific determinism but, on the contrary, absolute chance and contingency. From Bergson's premise of an individual pure perception contemplating the relation of coexisting image systems, Spinoza too would become a radical indeterminist, for he would be deprived of the generalized causality principle that animates his philosophy.

Moreover, the attempt to distinguish action from a passive or determined affection only nominally differentiates Bergson's thought, since for Spinoza affect can be both active and passive. Underneath these terminological differences, Bergson and Spinoza converge on the body as the limit condition of the mind. Through its very autological positing, or its *conatus*, the body acts into the world and strives toward increasing its power of acting. In Bergson's terms, the body functions as the constantly moving distinction between the two coexisting systems of images. Lodged within this double movement, as though in an electrical circuit, the individually perceiving body is at once active transmitter and passive conductor: "It is then the place of passage of the movements received and thrown back, a hyphen, a connecting link between the things which act upon me and the things upon which I act—the seat, in a word, of the sensori-motor phenomenon."[32]

With his reorientation toward the prerogative of action, then, Bergson is not distancing himself from Spinoza but more importantly positioning himself against a hegemonic current of Western thought, running through Aristotle as well as Descartes and Kant, that posits rational knowledge for its own sake as a distinguishing trait of human beings. In much scientific writing, too, the quest for truth is aligned with the greatest achievement of humanity, an Enlightenment conception par excellence. Critiquing this stubborn ideal as misguided, Bergson denounces philosophical and scientific conceptions like Newton's and Kant's, in which space and time are constituted by something like pure reason:

> Homogenous space and homogenous time are then neither properties of things nor essential conditions of our faculty of knowing them: they express, in an abstract form, the double work of solidification and of division which we effect on the moving continuity of the real in order to obtain there a fulcrum for our action, in order to fix within it starting-points for our operation, in short, to introduce into it real changes. They are the diagrammatic design of our eventual action upon matter.[33]

Concepts like the universe, in other words, enable us to immobilize nature so as to increase our mobility, our capacity for acting upon and within it. In Bergson's view, the mind simply furnishes the concepts most useful for bodily action.

By implication, the prerogative of action means that knowledge has to be considered not in itself but rather within a political field, in which any body and all bodies are necessarily engaged through the autological mediation of interest. The order of discrete, identifiable facts is equally conditioned by this mediation. As Bergson puts it, "that which is commonly called a fact is not reality as it appears to immediate intuition, but an adaptation of the real to the interests of practice and to the exigencies of social life."[34] In this sense, Bergson's philosophy, much like Spinoza's, implies a critical view of how the modern sciences are traditionally conceived. By refusing to accept an epistemological status for modern science that divides it from ontological concerns, both Bergson and Spinoza constitute significant undercurrents in modern thought that connect them to later thinkers such as Gilles Deleuze, Michel Serres, Isabelle Stengers, and Bruno Latour. Metaphysically, both Spinoza and Bergson open up to fundamentally political considerations of scientific practice by developing a perspective on the modern sciences in terms of their action in the world rather than their knowledge claims about it.

Yet neither Spinoza nor Bergson goes further in this direction. On the contrary, their key works unite in a concern for the potential reconciliation of metaphysical and scientific understanding. In turn, this raises questions about the status of the position from which they philosophize. Are we to understand metaphysics too under the predominant sign of action—or as distinct from the kind of practical knowledge the sciences offer? It seems we either have to ask about the "usefulness" of metaphysics—or we have to posit a purpose for metaphysics that differentiates it from scientific knowledge altogether.

Toward the end of *Matter and Memory*, a rather lyrical Bergson attempts to articulate the close parallels between contemporary physics, mathematics, and metaphysics, which in turn suggests a set of complementary practices for philosophy and science into the twentieth century.

> When we have placed ourselves at what we have called the turn of experience, when we have profited by the faint light which, illuminating the passage from the immediate to the useful, marks the dawn of our human experience, there still remains to be reconstituted, with the infinitely small elements which we thus perceive of the real curve, the curve itself stretching out into the darkness behind them. In this sense, the task of the philosopher, as we understand it, closely resembles that of the mathematician who determines a

function by starting from the differential. The final effort of philosophical research is a true work of integration.[35]

True integration—founded on difference, coexisting in deep admiration and respect for the universal continuity of becoming. Moving far beyond the lonesome ethical prescriptions of Spinoza, Bergson offered the new century a compelling, if not magnificent, vision of harmony that succeeded in moving exalted crowds into lecture halls, engaging masses of readers, interesting philosophers and scientists from all over the continent—all by invoking the ideals of comprehension and coherence. With Bergson, it seemed, the chaotic modern world was finally making sense again.

And yet within merely a few decades, the relationship between philosophy and science had never been more thoroughly bifurcated and antagonistic—and the virtual image of philosophy itself never more splintered. For just as Bergson authored a cogent account of contemporary physics, a quiet, century-long revolution was coming to the fore. In a span of merely three years after Bergson's 1896 publication, new experimental phenomena would boggle prevailing physical ideas: x-rays, radioactivity, electrons, canal rays. The attempts to render these phenomena coherent with physical explanation would significantly alter the direction of physics in the twentieth century, effectively overturning Bergson's conception of universal continuity, and lead toward an entirely new metaphysics.

In the end, this metalogical shift of physics would turn Bergson's Herculean efforts into a Sisyphean task. In this sense, I read his work as a harbinger for the fate of metaphysics in the twentieth century—in which the modern world-picture reaches its limit and doubles on itself.

PLANCK'S DISCONTINUITY

Outside the diverging discourses of physics and philosophy, the idea of entropy tending toward a maximum appeared to be more than a limited theoretical construct. Along the same exponential vectors as population curves, the physical conceptualization of energy in the nineteenth century correlates with its increasing mobilization: electricity became widely available and ushered in a new carbon era through the rapacious consumption of coal. By vastly transforming the human condition at the expense of a nature from which it had tried to differentiate itself, modern scientific culture expressed its profound inner contradiction in how energy was mobilized—and this contradiction could also be found in energy physics from its inception.

Energy is the idea underlying two pivotal constructs that together form the basis for the discourse of theoretical physics. One, as we have seen, was Maxwell's electromagnetic field theory—and the other was Maxwell and Boltzmann's kinetic theory of gases. In retrospect, these two theories turn out to be metaphysically irreconcilable because they rely on two different logics, expressed through different kinds of mathematics. Thus, the metalogical rise of theoretical physics is predicated on an internal bifurcation of energy—a precursor to the later incompatibility between general relativity and quantum mechanics.

In 1873, Maxwell mused on a contrast between these "two kinds of knowledge" in physics, one of which was dynamical, the other statistical and "belonging to a different department of knowledge from the exact science."[36] Specifically, Maxwell concluded, the laws of thermodynamics were statistical in nature—thus not absolutely certain, but only "morally certain." The most popular illustration of this insight is now called "Maxwell's demon"—a thought experiment that situates an observer (the demon) between two connected gas containers, able to follow the individual causal distribution of molecules as they flow from one container to another, from hot to cold, in the direction stipulated by the law of entropy.[37] From the perspective of this demon, it is impossible, Maxwell concluded, to ascertain that some particles would not be able to move the opposite way. The best that could be said for the laws of thermodynamics is therefore that they are highly probable with a degree of uncertainty. What statistical laws tell us is the inclination of a phenomenon to occur, not the causality by which it actually happens.

In the late nineteenth century, Maxwell's logical distinction was not commonly accepted. One physicist who remained unconvinced was the Austrian Ludwig Boltzmann. After years of conducting significant gas experiments that combined statistical and dynamical modes of calculation, he continued to insist that his H-theorem from these experiments proved irreversibility as an absolute, dynamical fact of nature.

Decades later he would concede the theorem was "merely" statistical, but he kept minimizing the importance of this logical difference, insisting that his proof was more than probabilistic—it was, he wrote, "extremely probable" and the chance it could be wrong was "extremely small," as though a logical gap could be bridged by scale.[38] For years Boltzmann put his physical work aside and embarked on a lengthy personal study of philosophy to find ground for his claim to a fundamental law of nature, only to be frustrated by its futility. In 1905 he wrote poignantly to the philosophy professor Franz Brentano: "Shouldn't the irresistible urge to philosophize be compared to the vomiting caused by migraines, in that something is trying to struggle out even though there's nothing there?"[39] Boltzmann simply could not understand what he was missing.

Another confused skeptic was the German physicist Max Planck. Throughout the 1890s, he worked on a theory of black-body radiation, which in effect combines three dimensions of Maxwell's work: electromagnetics, thermodynamics, and statistical mechanics. The physical parallel between kinetic gas theory and electromagnetic radiation was already established. Considered as systems, both move irreversibly toward equilibrium, defined in terms of Maxwell's statistical distribution, which could be applied to wavelength size rather than molecules. Planck was a rather conservative respondent to the new, progressive wave of energy-based physics. Like Boltzmann, he was concerned with reconciling the laws of thermodynamics with classical dynamics. To this end, Planck retraced many of Boltzmann's mathematical steps, year after year, eventually leading him in 1899 to the same frustrated conclusion as his Austrian peer. As the American historian of science Thomas Kuhn writes: "Both men had initially sought a deterministic demonstration of irreversibility; both had been forced to settle for a statistical proof; and both had finally recognized that even that method of derivation required recourse to a special hypothesis about nature."[40]

Only after 1910 would the statistical understanding of thermodynamics become firmly established in physics discourse, and this around the same time as probability became an autonomous branch of mathematics.[41] The slow recognition of the metalogical revolution mirrors the century-long historical rise of probability, in which statistical reason becomes increasingly associated with a philosophy of indeterminism, differentiated from mechanist deterministic laws.[42] Yet for the next century, this differentiation would be as unclear in scientific practice as in philosophy. Even if statistical laws were regarded as autonomous, the status of thermodynamics was paradoxical. On the one hand, its laws were developed from observable physical conditions of actual phenomena—in contrast to the laws of Newtonian dynamics, which were abstract and mathematical. Empirically, nothing appears more certain than the clearly visible fact that heat always flows from a hotter to a colder object. But on the other hand, Maxwell was arguing—and his successors would corroborate his point—that this phenomenal, irreversible reality is the one that could not be stated with certainty. To this day, as Cartwright and Stengers have explained in more detail, theoretical physicists are trained to accept that the laws of mechanics are primary to the "merely phenomenological" laws of thermodynamics, even if it is only the latter that can be witnessed directly in the laboratory. This devolution of thermodynamics to a second-class natural law is a common point of critique against a discipline that in the nineteenth century appeared to abdicate its physical common sense to mathematics.[43]

Bergson was among the thinkers who saw in the irreversibility of entropy the physical correlate of time's arrow. Moreover, because the difference between mechanical and statistical explanation implies indeterminism, it also leaves a loophole in physical description for the notion of human freedom. In fact, in the same 1873 paper that distinguishes dynamics and statistics as two kinds of knowledge, Maxwell also explicitly regards free will as a possibility within the framework of statistical knowledge. The probabilistic description of thermodynamics, in other words, appears to right the wrongs of Newtonian dynamics and leave the door open to human freedom. And in turn, much like Bergson's sentiment in *Matter and Memory*, such a positive recognition would harmonize physics and metaphysics.

Alas, the metalogical character of probabilistic thought is not so easily aligned with the prevailing system of either science or metaphysics, because the relation between them is logically irreconcilable. Maxwell's demon is not a real physical example of potential freedom to defy mechanist laws but an illustration of a logical limit condition. The irreversibility of statistical mechanics and thermodynamics is not a positive statement of time but, strictly speaking, only a function of probability. On the one hand, the classical order of dynamics is reversible because it is predicated on an inversion of autological causality, meaning that every cause is identified with an effect: under the principle of identity, A equals A, whose direct correspondence is the necessary condition for moving backward and forward between cause and effect without fault. Thus, the reversibility of rational mechanics is a function of its logical matrix. On the other hand, the metalogical order of statistical reason is predicated on the exclusion of autological causality, and thus relates to dynamics only by correlation. This lack of causal identity is why probability is nonreversible. Metalogically, a kinetic gas system, for example, is not irreversible because of the passage of time but because its mode of explanation is noncausal. In itself, then, the metalogical takes no positive account of time, freedom, or becoming. Only by differentiation from the established conceptual order of reversibility does it appear to take on positive meaning, precisely as a measure of that which cannot be circumscribed. In turn, this difference all too easily becomes hypostasized (by the principle of identity) and employed in the service of attempts to fill the gaping absence of causal reason: an ostensible proof of time, of human freedom, subjectivity, and so on. The misunderstood distinction offers a loophole for hopeful conjecture.

As a scientist in the classical Newtonian and natural theology tradition, Maxwell contributed to both a dynamical and a statistical conception of energy, without ever trying to resolve their incompatibility. For him, metaphysical reconciliation would necessarily involve the transcendental order of God, not the mathematical universe of physics. Therefore, he limited his claim to epistemology. Yet he clearly understood that the difference between dynamical and statistical explanation implied the difference between a fundamentally continuous and discontinuous universe. In a hypothetical turn of phrase, he described the consequence of statistical reason this way: "If the molecular theory of the constitution of bodies is true, all our knowledge of matter is of the statistical kind."[44] For both Maxwell and Boltzmann, the kinetic theory of gases was still limited to this untestable dependency clause: *if* matter is particulate. Certainly, kinetic theory lent some credibility to atomism, since it made individualized particles conceptually useful. But the actual constitution of matter was still far beyond the threshold of optical resolution. And more problematically, in Boltzmann's hypothesis the actual size of the atoms was arbitrarily chosen and could vary in relation to the other mathematical variables. From the perspective of a coherent atomism, this was contradictory. But how could we expect an autonomous statistical logic to be combined with mechanist logic without resulting in contradiction?

Eventually, Planck would find a roundabout way to a mathematical solution by inverting these contradictory logics into a case of mutual constitution within the same conceptual framework. After years of futile research on irreversibility, a disappointed Planck shifted his focus toward deriving from the black-body spectrum a radiation law that would hold up to experimental results. A black-body is a hypothetical space that functions much like Galileo's void, only in inversed form: it is not empty, but full, a plenitude of electromagnetic radiation.

> If a cavity with perfectly absorbing (i.e., black) walls is maintained at a fixed temperature T, its interior will be filled with radiant energy of all wavelengths. If that radiation is in equilibrium, both within the cavity and with its walls, then the rate at which energy is radiated across any surface or unit area is independent of the position and orientation of that surface.[45]

The black-body is a classic theoretico-experimental construct: if radiation in a specific thermal state is in equilibrium, then any local, mediating conditions can be ignored. In this sense, the formal problem is to derive a radiation formula in such a way that it can be extended to universality. By isolating local conditions and circumscribing temperature variation to a single state (T), the question becomes: How does the wavelength of a heated body change with temperature?

This formal similarity to the kinetic gas problem allows Planck to replicate Boltzmann's mathematical approach under the assumption that black-body radiation is a system like any other. However, this assumption conceals a crucial metaphysical difference. Whereas

Boltzmann operates with a distinctly bounded physical phenomenon—gas in a container—Planck is theorizing abstractly toward the limit condition of physics itself. He is trying to derive a mathematical expression for a compound foundational concept: energy, radiation, light, and electromagnetism. To make his problem conceivable mathematically, Planck finds it necessary to invent an additional hypothetical premise: a set of what he calls "resonators" that would absorb energy from, or be sensitive to, each wavelength frequency. The mathematical problem then becomes to calculate the frequency distribution of energy within this hypothetical space in terms of such hypothetical individual resonators. Because metalogical reasoning requires individual markers, Planck subdivides the energy continuum into elements of finite size, and he explicitly does this only for purposes of calculation. Crucially, he also reverses the order of the problem. Instead of stipulating suitable initial conditions that lead toward equilibrium, as Boltzmann did, Planck begins by assuming equilibrium in order to find the initial conditions, which in this case corresponds to the relationship between (hypothetical) resonators and wavelength. And unlike Boltzmann's variable-size atoms, Planck eventually finds that in order to make his energy elements correspond to the resonators, they have to be of a fixed size. Without this fixed discontinuity, the interaction between radiation and resonator would lead to an increasing dominance of oscillations over a diminishing frequency within the radiation field—that is, a runaway feedback effect that undermines any semblance of a mathematical solution. Hence, to preserve the statical equilibrium necessary to extend the problem to universality, any change in energy within the radiation spectrum would have to be expressed as multiples of the energy element e, in discontinuous jumps rather than continuous alteration. For Planck and the first physicists taking up his argument, there is no discernible physical reason why this should be so—it just happens to make the overall calculations cohere with experimental results.

Thus, in a new logical feedback cycle, two irreconcilable logics mutually constitute one another. In order to render the discrepancy between electromagnetic theory and actual experimental results coherent within a universal framework, Planck seeks recourse to a statistical, metalogical method whose sufficient condition is a discontinuous quantity. When he reverses the problem and assumes statistical distribution as his given, he reaches the point at which the mathematical continuity of electromagnetic radiation breaks down—the limit condition of the energy spectrum—and this too shows up as a discrete, fixed quantity. Calculating forward and backward, Planck is unable to eradicate this strange mathematical implication—a discontinuity in the continuity of radiation.

Notably, Planck's own understanding of his work differs from the meaning that is later ascribed to his efforts. At first, Planck appears to think he has merely produced a curious ad hoc result to a specific problem that he still views in classical terms. But five years later, Albert Einstein and other physicists begin to consider Planck's theory significant in a different way. As Kuhn puts it, "Without apparently having intended to do so, Planck had produced a concrete quantitative link between electromagnetic theory, on the one hand, and the properties of electrons and atoms, on the other."[46] Only upon this mediation of interest by his peers does Planck begin to describe his result as nonclassical and change the terms of his discovery to reflect its new status. What had first been described as an "element," in analogy to chemistry and Boltzmann's gas theory, he now baptizes the "quantum of action"—"quantum" signifying discontinuity and separability—and the acoustic analogy of "resonators" is changed to the binary term "oscillators."[47]

Metaphysically, Planck had (unintentionally) invented a sufficient condition for physics to ground itself in the very discontinuity between its two irreconcilable logics. Einstein made this suggestion in the first of his famous 1905 papers, "On a Heuristic Point of View about the Creation and Conversion of Light." Einstein argues that some contemporary experimental "phenomena involving the emission or conversion of light can be better understood on the assumption that the energy of light is distributed discontinuously in space."[48] Since the assumption of discontinuous distribution of energy is already embedded in Maxwell's statistical law of distribution, which itself implies discontinuity, Einstein's suggestion is to reinterpret the discrepancy that is bound to result from the two kinds of knowledge (irreconcilable logics) in physics. His move allows for a conceptual understanding of the nature of light corresponding to metalogical constraints instead of autological (continuous) functions. He goes on to show mathematically how Planck's "elementary quanta" can be considered independent of the theory of black-body radiation as "light-particles." Einstein's paper, suggesting the new operative assumption that light is particulate, is considered the beginning of quantum theory.[49] Notably, it contains no positive ontological claim for the discontinuity of light, but rather a strategic, operational reorientation of the framework within which light can be understood. In subsequent work, Einstein extended Planck's results to establish stronger coherence with prevailing theoretical models and allow for testable predictions.

Thus the quantum became intertwined with atomism, and what remained was to forge the experimental link between mathematical discontinuity and physical particle. Henceforth, the progress of physics was a matter of folding metalogical reasoning back into an identifiable framework—reconstituting metalogical assumption as experimental fact.

THE EMERGENCE OF PARTICLE PHYSICS

Uncoincidentally, the first modern particle discovery happened in an experimental vacuum. In 1897 the electron was derived from experiments with electric discharges in gases inside vacuum tubes. In these specially designed tubes, it was known that invisible cathode rays would light up upon contact with certain residue materials. The British physicist J. J. Thomson was looking to prove what he believed, that matter had a molecular structure. He found a way to deflect the rays in the tube, and by differentiating electric and magnetic interference he could measure its ratio of mass to charge. According to Maxwell's theory, electromagnetic waves do not carry a charge, and Thomson's result therefore indicated a discrepancy in the prevailing model. Framing his mathematical result in a different model, developed by Hendrik Lorenz a few years earlier, Thomson hypothesized the existence of an isolated entity. He called this new being a corpuscle, partly because he thought his particle was different from other electrons that had already been theorized (but not produced empirically), and partly for its classical metaphysical connotations. Thomson eagerly pronounced the implications for atomic theory:

We have in the cathode rays matter in a new state, a state in which the subdivision of matter is carried very much further than in the ordinary gaseous state: a state in which all matter . . . is of one and the same kind; this matter being the substance from which all the chemical elements are built up.[50]

Thomson claimed, in other words, to have discovered substance itself in a test tube.

Alas, things were not that simple. In a case study of the electron, Brigitte Falkenburg analyzes the complicated intertwining of theory and experiment involved in Thomson's claim. As she points out, Thomson's "measurement did not in any way test the hypothesis that cathode rays consist of single massive charged particles. It only confirmed a consequence of this hypothesis," expressed in terms of Lorentz's mathematical model, which itself already implied atomism. None of the properties Thomson ascribes to the particle—inertial mass, electric charge, point-like or local behavior, and a trajectory through classical space-time—was actually measured by him, nor are they subject to independent measurement. Rather, as Falkenburg puts it, "they are connected to each other by classical dynamics" in the same tacit particle concept.[51] By experimentally measuring the ratio of mass to charge in the observed phenomenon, Thomson really projected a dynamic structure onto his result.

As the science historian Helge Kragh suggests, Thomson is celebrated as the discoverer of the electron less on account of his specific experimental findings than for the boldness of his claim, which succeeded in attracting the interest of contemporary physicists. Within the next few years, other isolated experimental phenomena with approximately the same mass-to-charge ratio were found in experiments of photoelectricity, beta radioactivity, and thermionics, and a similar quantity could be inferred from other fields of physics. For believers in atomism, the mathematical stability of these results was strong indication of a discontinuous, identifiable phenomenon. But the ontological status of the electron, because it lay far below the optical threshold, would be impossible to discern from experiment alone. By the 1920s, the new paradigm of quantum mechanics would hold that electrons could be measured as both a particle and a wave and that its "true" nature was related to how the experiments made it appear. As recently as 2006, precision experiments in high-energy physics concluded that the electron is a so-called point particle—it possesses no structure beyond what the theories of quantum mechanics and relativity demand for their conceptual coherence.[52]

Without a microscope powerful enough, how can we "see" an atom? In the experimental search for hypothetical atomic properties, new radiation phenomena were mobilized to the cause. Most instrumental of these were so-called scattering experiments, which measure the interaction of radiation and matter by counting the relative frequency of "particles" bouncing at specified angles. Early scattering experiments are direct ancestors of accelerators such as the Large Hadron Collider, in which particles are presupposed as part of the probabilistic formula of measurement. Initially, they relied on kinetic theory, wherein energy is assumed to be distributed in particulate form. Gradually, the experiments would grow to encompass more complicated statistical models that derive individual particle predictions from the observable level of collective effects. In this tautological sense, scattering presupposes the very particularity it is designed to measure.

Moreover, scattering was the basis for the metalogical leap beyond Abbe's optical limit into what was known as the electron microscope. This third historical jump in resolution occurs with the first prototypes in the early 1930s. Instead of using natural or electric light, the electron microscope is a closed system that focuses a beam of electrons within a tiny wavelength, thereby achieving, according to Abbe's formula, a theoretical limit of resolution close to one hundred times the maximum of optical light microscopes. By simultaneously extending the mathematical framework of optics and breaking free from its limit, the electron microscope is a paradoxical construction that exploits both dimensions of the wave-particle duality in quantum physics. Insofar as electrons can be considered waveforms, it functions in direct analogy with optics as an extension of microscopy, on the same logical plane. However, insofar as electrons are considered particles, the microscope image is not analogous to the continuous contrast of light absorption—rather, it is produced by the differential scattering of electrons.[53] In this sense, whereas the precise function of Abbe's optical light microscope is circumscribed by a mathematical framework, the electron microscope is mathematical in its very operation.

Nevertheless, despite the great magnifying leap of electron microscopy into a probabilistic universe, it remains inadequate for imaging at an atomic level. For almost a century, atoms have been known and taught as the building blocks of the universe, but to this day, nobody has ever seen an atom in any empirical sense. Does this mean the theory is false?

The answer depends on what is warranted by a scientific method of demonstration, and the goalposts for what counts as discovery were significantly moved in the course of the twentieth century. The claims to have discovered particles in each case involve a complicated logic. As an example, by the mid-1920s the light-particle was submitted to scattering, rebaptized the "photon," and considered empirically verified in a coordination of theoretical models and measurement similar to the electron's. In turn, this experimental confirmation of the light quantum would lend further instrumental support to the atomic models of quantum mechanics. Thus, a paradoxical doubling of discontinuous inventions reverberates in a mix of dynamical and statistical models.[54] The particle concept is mutually constituted by, on the one hand, the quantum—the limit condition of light phenomena—and on the other hand by the atom—the limit condition of matter. Both of these inventions derive from the same metalogical operation, and their independent reality status is implied by their mutual conceptual unity. In this precise sense, the predominant physical and metaphysical problem of the nineteenth century, the relation between energy and matter, finds its resolution in an experimental invention of particularity. And the crux of the invention is not that it is true or false but that it, insofar as it became accepted practice, allowed physicists an entirely new range of operation, new modes of action, new forms of research.

In *Matter and Memory*, Bergson had postulated as axiomatic to his intuitive method that "all division of matter into independent bodies with absolutely determined outlines is an artificial division."[55] To speak of particularity in the manner of twentieth-century physics would for Bergson signify a mental operation, a "spatialization of thought," that lacks temporal reality. After the metalogical turn, any such attempted distinction between real and artificial would become hopelessly futile. In fact, as Falkenburg also points out, the fundamental modern philosophical differentiation between ontology and epistemology, being and knowing, is thoroughly incapable of illuminating the development of early-twentieth-century physics. Is the electron (or the atom, or the light quantum) ontologically real or an epistemological construct? This can be fodder for endless academic debates, but first and foremost, the modern particle concept is theoretically and experimentally *operational*. The particle is the condition of possibility for physicists to act upon and into matter in such a way as to extend their research beyond the limit of optical instruments. In this sense, particles are mobilized into being. As Falkenburg describes it, "the only decisive proof of particles is apparently to *make* them and to *use* them as tools in other experiments."[56] Even as the technological means of extension fail to reveal their certain existence, and in fact point to the opposite conclusion, particles remain a necessary condition for research, and therefore physicists act as if they exist.

In other words, the modern particle—be it in the form of the electron, atom, quark, or quantum—is neither ontological nor epistemological. It is metalogical. It is real because we can use it and make more of it, which begets more in turn.[57] Probability begets particularity begets proliferation. And among the most significant consequences of the invention of particularity, as the next chapter addresses, was the new basis for a science of cosmology.

Bergson, despite his remarkable insight into the sciences of his day, never foresaw this drastic development in physics, and neither did he foresee the unprecedented changes in his own culture. Within the first three decades of the twentieth century, in the viral growth of society and in the wake of the cataclysmic Great War, vertiginous reaction formations were alternately turning against and clinging onto its own cultural foundations, while being mobilized anew for even more unfathomable modes of catastrophe. In this volatile cocktail, nuclear research would play a particularly devastating role. Soon, Bergson's harmonic philosophy of continuity would be quashed by the explosive growth lines of a culture proliferating beyond any sense of rational human control. Moreover, the metalogical revolution that transformed the sciences in the late nineteenth and early twentieth centuries had a correlated effect in a wider chasm opening between the discourses of physics and philosophy. A generation of physicists trained in classical philosophy and philosophers trained in the general sciences would soon give way to a professionalized bifurcation between increasingly specialized disciplines.

In this context, Henri Bergson and Albert Einstein both exemplify the passage of a last generation of metaphysicians who could still conceive of their work in the vein of seventeenth-century natural philosophy. In 1922, Bergson and Einstein, both intellectual celebrities of their time, met for a public debate. Although the theme of the debate was the meaning of time and relativity, a deeper and more controversial subtext soon surfaced: Who can most authoritatively determine metaphysical questions—physicists or philosophers?[58]

During the debate, Bergson argued that philosophy still played a crucial role in understanding the nature of time. But Einstein was not willing to cede ground and posited a full metaphysical reversal. Bergson had claimed in *Matter and Memory* that the abstract physical conception of time failed to account for its durational, autological reality. Now Einstein countered that the absolute reality of physical time is in fact different from what he called the "time of philosophers," which is "nothing more than mental constructs, logical entities."[59] Mathematical physics, Einstein argued, actually has the firmer grasp on reality, while philosophers deal with artificial conceptual constructs. In effect, Einstein declared metaphysical independence from physics. However unfairly it arose, the widespread view in the aftermath was that Einstein had won the debate because Bergson failed to recognize the independent reality of physical time. As Gilles Deleuze would later put it, Bergson's intervention "led to so much misunderstanding because it was thought that Bergson was seeking to refute or correct Einstein, while in fact he wanted . . . to give the theory of relativity the metaphysics it lacked."[60] Bergson always pleaded for a complementary role for philosophy, but his modest proposal was squarely rebuffed. In this sense, the encounter between Bergson and Einstein marks not only the waning of Bergson's influence in his own time but, more profoundly, the historical passing of philosophy into increasing irrelevance.

Certainly, there was no shortage of attempts at cosmic syntheses in this era, Kantian or otherwise. New philosophical currents, most notably phenomenology, arose with the aim of recovering some sense of the lost autological "life-world." In turn, these movements faced ever starker opposition to the positivist movements that wanted to abolish metaphysics from philosophy altogether. By the time the young German philosopher Martin Heidegger waded into the bitterly fraught scene in Europe, the vicissitudes of Bergsonism had revealed the incipient resentment of modern mass popularity by mercilessly turning the French philosopher from cause célèbre to cliché. As Whitehead once remarked, "a system of philosophy is never refuted; it is only abandoned"—and Bergson's fate during the world war decades illustrates it all too well. Although perhaps no thinker had done more than Bergson to advance a philosophy of time, life, difference, and experience in distinction from prevailing positivist and neo-Kantian currents, the young, up-and-coming Heidegger made scant reference to Bergson in his own work. In a dismissive one-paragraph analysis, Heidegger first reduces Bergson's thought to simply a reversal of the Aristotelian conception of time, and then, as if that were not enough, claims that Bergson misunderstands Aristotle anyway.

Thus, the Frenchman who eagerly staked out a new orientation for a philosophy of life, drawing on the heretical Spinoza, was recast as yet another culprit in the history of thought from which the young German phenomenologist, himself caught in the increasingly fashion-cyclical vortex of professional philosophy, eagerly sought to differentiate himself. Although Heidegger would be much concerned with the development of physics, his stance implied he would always speak from the other side of a widening chasm. Heidegger could bemoan and critique the technological advances of his time, but he would never address it in terms that a new order of physicists could understand.

As a consequence of the metalogical revolution, the new physics that emerged in the twentieth century became an independent force, unshackled not only from philosophy but also from its own metaphysical foundations. It reached into the future as it disengaged with its past, forging a new macro-cosmology from inventions in micro-physics. In the new metalogical order after Einstein, physics would venture far beyond the limits of the given, make discontinuity its new metaphysical foundation, and discontinue its own metaphysical tradition.

4

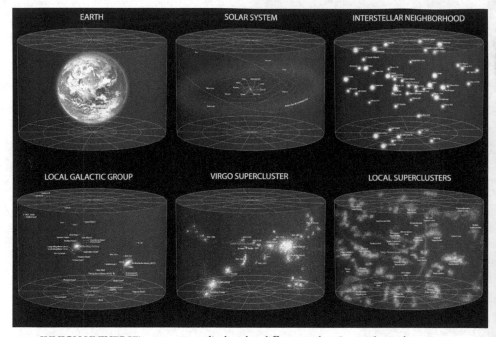

WHICH UNIVERSE?: our cosmos displayed at different scales. Copyright Andrew Z. Colvin, Creative Commons.

METAPHYSICS WITH A BIG BANG

The Invention of Scientific Cosmology

Over the entrance to the gates of science's temple are written the words:
Ye must have faith.

—*Max Planck*

HYPOLOGICAL ORIGINS

In 2009, a large rocket was launched into orbit from French Guiana carrying the Herschel Space Observatory, developed by the European Space Agency. By 2013 the Herschel spacecraft had run out of coolant liquid, as expected, and was left in space to die. Its carcass currently hovers around the so-called L2 point, 1.5 million kilometers from Earth, where it is soon to be accompanied by the brand-new James Webb Space Telescope (JWST), which is estimated to last up to five years in orbit.

Herschel was the third in a series of astronautic contraptions designed to measure in the infrared light range, a capability that the JSWT is set to expand. Unlike a particle accelerator, which has integrated mathematical discontinuity into its reconstitution of microphysical events, the Herschel and Webb telescopes function according to infrared wavelengths, which has the capacity to make readings through clutter and debris that blurs the distant skies. Similarly to the LHC, the interface of the operation belies the conceptual simplicity of "an observing telescope." Consisting of hundreds of individual digital shutters, the Herschel and the JWST "observe" in the sense that they produce constant streams of data that are later reconstituted into imagery according to a multitude of parameters. After all, photographic images of extremely distant blinking lights, blurry dots, dust balls, and random objects provide no clues to the content of the universe; therefore, cosmologists require at every step of the process an alliance with a map, a theoretical framework, in order to determine what is useful to their research. Lodged in a space too far away for obvious means of falsification, this frontier science is therefore haunted by a kind of circular logic: in concordance with the predictions of the big bang theory, the telescope searches only in those specific wave regions where it might find evidence that vindicates the theory. And if it doesn't, the parameters can later be adjusted to make the "observations" fit. However, if the search actually succeeds in vindicating the predicted model completely, it becomes difficult to claim the need for more research. Therefore, the best-case scenario for the JWST, as with the LHC and similar mega-experiments, is that it appears to affirm the theoretical framework while presenting some surprising findings or unforeseen complications that requires new scientific experiments.

In the age of "Big Science," the primary purpose of research is to produce more research. What did the Herschel accomplish in its three-year life span? In the proud summary of the European Space Agency, Herschel made over 35,000 scientific observations, amassed more than 25,000 labor hours' worth of science data from about 600 different observing programs, plus a further 2,000 hours of calibration observations. And with a price tag of $1.4 billion, Herschel cost merely a sixth of the JWST. Speaking to *Universe Today*, Herschel program manager Thomas Passvogel noted: "Herschel's ground-breaking scientific haul is in no little part down to the excellent work done by European industry, institutions and academia in developing, building and operating the observatory and its instruments."[1] It is no coincidence that the scale of the current cosmic operation first and foremost is expressed in the form of economic figures. Driving large-scale technological innovation and industrial output has become an end in itself.

The European-based Herschel and the U.S.-led JWST are recent representatives of a giant industrial leap that scientists like Einstein probably never would have predicted. The new face of modern physics and cosmology that developed in the twentieth century rode on the back of enormous military-industrial budget increases since World War II. Through most of the postwar era, federal U.S. money was synonymous with military money, a vast capital flow that funded a new and unprecedented metalogical expansion of research in a new giant constellation of interests between science, government, military, and industry. In the period between 1938 and 1953, federal funds for basic science research, adjusted for inflation, increased by 2,000 percent. And through the 1950s budgets were maintained, leveling off through the 1960s and 1970s and rising again in the 1980s under Reagan. Physics found a new role in the twentieth century as a war industry that would change its mandate forever.[2]

Along with new scales came new scopes. The new theoretical physics of the postwar era emerged with different values, uninhibited by metaphysical constraints. As Stengers argues, "The new identity of physics passionately requires that the world justify, and that humans accept, physicists' right to freely negotiate how and to what extent obligations bind them to the phenomena beyond which their vocation demands they venture. . . . It reflects the physicist's freedom not to take 'observable phenomena' into account but to *discount* them."[3] The metaphysical freedom that Einstein had fought for in the early twentieth century would come to mean a freedom to cherry-pick observable phenomena that fit operational models and a freedom to move back and forth freely between empirical and theoretical modes in reciprocal determination. Physics had finally achieved Plato's ideal of the philosopher-scientist.

This final chapter is about the twentieth-century emergence of cosmology as a big scientific discipline—how its joint venture with physics after World War II provided us with a new story about the universe, a new creation and understanding of space and time.

Like all new beginnings, the reinvented physics emerged like the two faces of Janus. As the imperial Roman god of beginnings, gates, doorways, the god for whom our first calendar month is named, Janus transcends history by looking in two temporal trajectories at once, in a singular moment that is always between past and future. For Hannah Arendt, writing in the 1950s in America, the two-faced interval between past and future indicates not the continuing present of a linear time but rather a volatile moment of thought itself, the gap in which thinking occurs, pressured between temporal forces on two sides. Historically, she writes, such an odd in-between period in our culture inserts itself from time to time, "when not only the later historians but the actors and witnesses, the living themselves, become aware of an interval in time which is altogether determined by things that are no longer and by things that are not yet."[4]

The no longer and the not yet: this is the inversion of what philosophers after Hegel called the "'always already." The no longer marks a rupture in an ostensibly cyclical time, but it does not yet appear clearly to the actors and witnesses of history. The "living themselves" recognize in their thinking that the past is no longer, that something is changing, but they are not yet able to determine how. In this moment between past and future, the retroactive unfolding of history, in which what is will always already be, has yet to take place.

Arendt's words presciently describe their own historical moment of utterance. In the wake of two calamitous world wars, yet barely at the beginning of that extensive technological escalation to be known as the Cold War, the 1950s was precisely such an odd historical interval between past and future. The cultural shock of an apocalyptic war machine ravaging the planet ensured fertile ground for new beginnings. This was no less true in science, where the grandest of all possible beginnings was about to be invented. Through a scientific battle of rivaling metaphysics inside and outside the burgeoning community of cosmologists, two distinct hypotheses of the universe were articulated and pitted against each other—the (losing) steady state theory and the (winning) big bang theory—and in the wake of the battle, the modern universe was reinvented as a scientific cosmogony.

What was new about the big bang? In the conception of the cosmos prevalent until Arendt's time, the world had neither a beginning nor an end. The modern calendar, established in the late eighteenth century, made the birth of Christ a neutral degree zero marker, which allows history to stretch infinitely forward and backward in time on scientific time scales of millions of years. But the new scientific universe that was emerging would once again, much like a Christian cosmology, have a beginning, a moment of creation, and therefore quite possibly an end—a finitude that could, with sufficient research investment, be calculated. It is a beginning that is mathematical and generated by a scientific discipline—but is it therefore scientific?

In this chapter, I will argue that the subject matter of cosmology exists so far away from the autological realm of the sensible and is so thoroughly determined by metalogical frameworks that it also required a renewed faith in the project of universal mathematics. In my reading, the big bang hypothesis did not win out because it is true in any positive, verifiable, and empirical sense but rather because it most effectively gathered and mobilized interest in the scientific community for its explication of a few fundamental constraints. Like the invention of the particle, discussed in the previous chapter, the hypothesis was so instrumental to further research that it soon became reality. Arendt offers a precise conception of this logic at work:

> What was originally nothing but a hypothesis, to be proved or disproved by actual facts, will in the course of consistent action always turn into a fact, never to be disproved. In other words, the axiom from which the deduction is started does not need to be, as traditional metaphysics and logic supposed, a self-evident truth; it does not have to tally at all with the facts as given in the objective world at the moment the action starts; the process of action, if it is consistent, will proceed to create a world in which the assumption becomes axiomatic and self-evident.[5]

A hypothesis is the conception of a beginning—an invented origin, in analogy to Janus, a first calendar month in a cyclical time. January is the beginning of the year, because through consistent action in our cultural history we have come to structure our lives in such ways as to make January the self-evident beginning of that natural cycle we call a year. There is no given reason in nature for this to be so. In fact, if we were looking for a more "natural" start of the year, it would occur around December 21, when the sun turns relative to Earth, or with the spring equinox, when the exact balance of day and night tilts in one direction (the start of the year in, e.g., the astrological and Balinese calendars). Rather, January is first because a Roman emperor once said so—and it has become so because we made it so, and we still do. Similarly, most national communities are founded on "origin stories" that give their members an actively cherished narrative of how their history began, creating a semblance of order from chaos. In this sense, the hypothesis constitutes a tacit framework within which the world can be actively re-invented and become historically true.[6] This, as Spinoza and Heidegger would readily recognize, is the logic of identity itself—how A becomes A, and thus the beginning of an alphabet.

This identity logic I propose to call *hypological*, using the Greek prefix for "under." Hypology means positing something as under itself, in the manner of a framework, established through the retroactive unfolding of a transformation that makes its terms appear self-evident. In this precise sense, hypology also constitutes a logic of "under-standing." To "under-stand" is to stand under conditions as they are enframed for me. That is, I understand when I actively participate in the idea that can order my experience in a certain way. I understand the concept of January as the first month by adhering to the cultural practice of, say, celebrating New Year's Eve.

Arendt first derived this logic from her study of totalitarian political regimes. That it should so accurately describe the history of twentieth-century cosmology is perhaps not entirely coincidental. For as this chapter will try to demonstrate, what is most fundamentally at stake in big bang cosmology is not the beginning of time or the fate of the universe but rather the origin story of our own culture and the survival of our metaphysical world-picture.

EINSTEIN'S GENERAL RELATIVITY

The new cosmology of the twentieth century expanded on two fronts. Because there were few telescopes to provide enough data, it crucially relied on micro-physics to get the macro-universe into theoretical view. Terrestrial experiments like the LHC could provide the theoretical basis for astronomical observations, and vice versa, as discussed in chapter 1. Theoretical physicists limited by the scope of accelerator technology are hopeful that new astronomical telescopes will return the favor and feed them with new data on cosmic phenomena that could lend support to their own theories.

This productive junction between the micro and the macro is made possible by the pivotal principle of accelerator technology, which is also at the heart of general relativity: the higher the energy or momentum, the lower the wavelength, that is, the deeper into the nucleus these microscopes are able to probe. Increase the electron voltage a thousand times and you will be able to see "nature" at a thousand-times-smaller wavelength. To produce a particle of a certain mass, in other words, a corresponding energy is required. As an experimental expression of this exact quantitative relationship—a relational constancy between speed, energy increase, and mass decrease—the particle accelerator and its link to the outer universe is therefore metaphysically rooted in perhaps the most famous physics equation of the twentieth century: $E = mc^2$.

Einstein's formulation of mass-energy equivalence is usually described as a revolutionary break with the classical Newtonian universe. Yet it would be more accurate to say that Einstein's theory of relativity is an extension of Newton's universe, insofar as it saves the classical universalist configuration in a new and more simplified expression. For even if its extensive field equations are forbiddingly complicated, and even if its exact implications are contested, the metaphysics of general relativity essentially emerges from a constellation of three hypothetical principles. The first two, combined in Einstein's special theory of relativity (1905), are the principle of relativity and the principle of constancy of the speed of light. The third, which enables the general theory of relativity (1917), is the principle of equivalence. While the constancy of the speed of light is a critical invention of Einstein's age, the other two are new versions of principles put forth by Galileo and Newton.

First, the principle of relativity. Galileo argued that the laws of physics are the same "in all inertial frames," that is, in all cases of uniform motion, in which the medium is circumscribed and excluded. As Einstein put it, "the laws of nature perceived by an observer are *independent* of his state of motion."[7] To physicists, "relativity" refers to the relative motion within which an observer is situated—one frame moving faster or slower than the other—and from which any chosen movement thus can be considered at rest in relative terms. The Newtonian system made inertia this absolute foundation for measuring relative movement, and the principle of relativity enabled the constitution of space as a system of coordinates in three continuous dimensions. But, argued Einstein, "this relativity had no role in building up the theory. One spoke of points of space, as instants of time, as if they were absolute realities."[8] The crux of the theory of special relativity was to extend Newton's principle to time itself, now conceived as a fourth dimension of a new metaphysical construction called space-time.

An immediate consequence of special relativity, which would be further extended in general relativity, was that space and time were no longer to be considered as real physical entities but rather as derivative geometrical functions of the new pivotal concept of modern physics, the event. "It is neither the point in space, nor the instant in time, at which something happens that has physical reality, but only the event itself."[9] This new physical conception of an event—Einstein's 1905 paper uses the word eleven times in a page and a half—thus functions as a bridge between two sets of symmetries embodied in the $E = mc^2$ equation—between matter / energy and the derivative space / time.

However, this symmetry hinges on a specific condition in which the relative measure of matter and energy is invariant for any frame of reference. For this reason, Einstein's preferred term for his invention was the "invariant theory," since "relativity theory," which was first used by Max Planck in 1906 and quickly became the accepted term, often leads to misunderstanding. To ensure the invariance between relative coordinates within the posited system, some constant reference was required in place of what in Newton's system had been absolute time. The new absolute of the theory of relativity, suggested by a few nineteenth-century experiments, is expressed by the letter c, for "celeritas"—the speed of light in a vacuum.

Einstein explains the principle of constancy this way:

> In order to give physical significance to the concept of time, processes of some kind are required which enable relations to be established between different places. It is immaterial what kind of processes one chooses for such a definition of time. It is advantageous, however, for the theory, to choose only those processes concerning which we know something certain. This holds for the propagation of light *in vacuo* in a higher degree than for any other process which could be considered.[10]

Einstein explicitly seeks recourse to a hypothetical concept that can maintain a high degree of certainty and yet be more specific than similarly hypothetical notions like Galileo's void or Planck's black-body. A vacuum in physics is sometimes called "free space," because it corresponds to conditions that do not exist in nature and can only be experimentally constructed.

Metaphysically, this means that the independent physical reality conceived under the principle of relativity is conditioned by an independent nonphysical reality, or physical non-reality, the concept of a vacuum. That which absolutely exists independent of local conditions (and independent of us) requires the guarantee of that which does not and cannot exist in the actual physical world, other than on exceptional experimental terms. The principle of constancy posits that, if the speed of light is the *same* under the *same* hypothetical conditions, energy and matter can be considered equivalent in the independent physical reality posited by the principle of relativity. This is how Einstein aligns two different logical planes: physical reality is independent (of us) because the speed of light in a vacuum is constant. The principle says, if something is constant within conditions guaranteed to be constant, that constancy can be said to be constant. It is so because it is so. But how did it come to be so?

The crucial next step that allows for this general symmetrical identity is Einstein's next invention, which hinges directly on the principle of constancy. The point, as Einstein makes explicit, is not what the constancy is or whether it actually exists, but rather what it *does*. In this case, the principle enables the mathematical construction of a new point of view through which the world can be reinvented. Einstein's reasoning is perfectly hypological. Whether or not the assumption of a constancy of the speed of light is an objective fact in the given world at the time of conception does not matter, for it will eventually, through consistent action, be turned into an axiomatic truth. Wherever we look in physics literature, we will not get further insight into the speed of light than that it is constant first and foremost because it is defined as constant. And from this foundation a whole process of scientific action in the twentieth century unfolded. With the scientific hegemony of relativity theory, both in its special and general versions, the hypothetical speed of light has become a new given, not least as the absolute measure of the internationally standardized metric system. One meter is today defined by the International System of Units as the distance traveled by light in free space in 1/299,792,458th of a second. In this sense, the modern structure of the world, its system of identity through which all its key metrics are expressed, is hypologically constituted.

Because it is not based on an actual phenomenon, this hypological framework has a clear theoretical upside, as Einstein recognized: conceptual stability. However, it also comes with a practical downside: cascading complications in the gap between theory and phenomena. Already half a century before Einstein's theory, it was well established that the speed of light is not constant but in fact varies with actual physical conditions. Air slows down light, water and glass even more so—a phenomenon known as refraction, which, to be sure, can be calculated, but again only with recourse to idealized conditions in the same hypological manner.

Moreover, actual physical conditions inevitably involve the autological force of gravity. Einstein had tried to avoid this with his special relativity theory, which is "special" in the sense of being restricted, in the use of both its principles. Whereas the principle of constancy is explicitly hypothetical, the principle of relativity is restricted to uniform, linear motion (inertia) and does not account for nonuniform motion, such as acceleration due to gravitation. When the theory is applied to specific conditions and recalculated for them, complications crop up that only rise with the scope of the problem. For Galileo, the discrepancy between theoretical void and actual conditions in his local rolling ball experiment could be minimized and controlled. Can the same thing be said for calculations for the entirety of the known universe?

This is where the third principle comes into play. As Einstein was looking for a way to generalize his special relativity theory, he once again turned to a postulate of Galilean-Newtonian mechanics in order to modify it for his own needs:

> The ratio of the masses of two bodies is defined in mechanics in two ways which differ from each other fundamentally; in the first place, as the reciprocal ratio of the accelerations which the same

motive force imparts to them (inert mass), and in the second place, as the ratio of the forces which act upon them in the same gravitational field (gravitational mass).[11]

These two concepts therefore appear in classical mechanics as asymmetrical, but for no reason that can be explained in terms of the phenomena themselves. Characteristically, Einstein develops thought experiments to argue for a perspective from which the difference could be perceived as symmetrical, and thus equalized. His most famous example goes like this: If you are standing inside an elevator in free space—that is, a hypothetical vacuum—which is lifted upward by the same speed as the force of the gravitational pull on Earth, you would not be able to distinguish between acceleration and gravitation. And inversely, if you stand in an elevator located on Earth that is in free fall, you would not feel the gravity that is now practically canceled out by its symmetrical opposite, the acceleration of the elevator. Never mind the absurdity of the example, argues Einstein, for it illustrates what he calls the principle of equivalence, which "signifies an extension of the principle of relativity to coordinate systems which are in non-uniform motion relatively to each other. In fact, through this conception we arrive at the unity of the nature of inertia and gravitation."[12]

It is the principle of equivalence that makes special relativity "general," which here means universal, because it enables an entire mathematical universe through a simple set of explanatory symmetries. In this sense, the principle of equivalence functions as a kind of generalized identity principle. That which can be posited here and now is the same as that which can be posited anywhere and anytime. In arguing for his general theory, Einstein demonstrates hypological reasoning of the shrewdest order, alternating between appeals to empirical facts and to hypothetical scenarios. In his thought experiment, he simultaneously argues from experience—the person in the elevator would not "feel" the difference of the motions—and hypothetically, based on the absurd premise of an elevator in a vacuum. As in the case of special relativity, what matters is not coherent reasoning but construction of a perspective that allows for an ingenious theoretical simplification. In one and the same operation, the principle of equivalence dispenses with the restriction to the inertial systems of special relativity and the persistent problem of gravity in classical physics. Newton's solution to this problem was to turn gravity into a universal constant of his cosmology, analogously to what Einstein does to the speed of light. Einstein's solution is to make gravity a geometrical spatiotemporal property of mass–energy equivalence itself. Again appealing to known experience, Einstein goes on to justify the metaphysical downgrading of inertial systems by arguing that they constitute a very limited case of actually known properties of the universe. In fact, he points out, inertia is itself a dubious concept:

> The weakness of the principle of inertia lies in this, that it involves an argument in a circle: a mass moves without acceleration if it is sufficiently far from other bodies; we know that it is sufficiently far from other bodies only by the fact that it moves without acceleration.[13]

In the Newtonian system, then, inertia provides a secure foundation insofar as it guarantees no external interaction. As Einstein himself explains it, inertia refers only to itself—it is tautological. But Einstein does not banish this tautological weakness from his own system. On the contrary, the upshot of the principle of equivalence is rather to extend this tautology to the non-uniform field of gravitational movement, which under the new regime directly governs the relationship between body and force in an interaction. The problem with such a conception is that when a body is itself involved with a force, feedback happens. As astronomer Lindley puts it:

> When two bodies are pulled apart against their gravitational attraction, energy must be expended, and if they come together energy is released; but energy, as Einstein so famously proved, is equivalent to mass, and mass is subject to gravity. Therefore, the energy involved in a gravitational interaction between bodies is itself subject to gravity. Gravity, if you like, gravitates.[14]

Herein lies one significant metaphysical consequence of Einstein's principle of equivalence, which implicitly assigns to force a self-reinforcing tendency. What does it mean that gravity gravitates?

The quantitative dimension of this problem has preoccupied many physicists. In the 1930s and 1940s, for example, the development of quantum mechanics was consistently stymied by the stubborn appearance of mathematical infinities whenever calculations of force were attempted. This problem was only formally circumvented in the 1950s with the invention of a pragmatic mathematical trick called "renormalization," in which the troublesome infinities were effectively canceled out of the calculations altogether. But while renormalization was a crucial step in the quantum field theories that ventured to unify the strong, weak, and electromagnetic forces, gravity persisted in complicating the picture. Due to its pervasive and attractive nature, whenever the gravitational force between two bodies is calculated, there is an immediate and recalcitrant feedback into the calculations that no established mathematical tricks have been able to cancel out. In this sense, the difference between gravity and the three other differentiated forces of today's Standard Model bespeaks the bifurcation at the heart of contemporary physics, which theoretical physicists and cosmologists aim to solve, between explanations on a quantum scale and those on the cosmological scale of general relativity.

Metaphysically, the problem that haunts modern physics is that both gravity and the speed of light are fundamental limit conditions of scientific experimentation. Considered separately, each constitutes an essential asymmetry in a universal cosmic symmetry: gravity refuses normal mathematical integration; the speed of light resists assimilation to actual physical conditions. Instead of considering this a problem, Einstein applied the same logic as Planck before him, turning a limit condition of experiments into conditions for mathematical operation. The problem, in other words, was turned into a hypology.

PLANCK'S CONSTANTS OF NATURE

In forging the constancy of the speed of light as a new standard, Einstein's theoretical pragmatism played right into the spirit of physics at the time, which was moving toward inventing new forms of constancy. Like quilting points, constants could hold the expanding fabric of theoretical physics in order. In this development, Planck's work in particular was crucial to the later emergence of modern cosmology and therefore warrants some explication.

In the late nineteenth century, the chaotic metastasis of Western culture increasingly demanded international orders of standardization, from the metric system to world time zones. Among physicists, too, different unit systems were being proposed to regularize their work. Planck believed deeply in the metaphysical independence of nature from human or cultural constructs, and he thought the problem with late-nineteenth-century measures of standardization was their sheer contingency.

> All the systems of units that have hitherto been employed . . . owe their origin to the coincidence of accidental circumstances, inasmuch as the choice of units lying at the base of every system has been made, not according to general points of view which would necessarily retain their importance for all place and all times, but essentially with reference to the special needs of our terrestrial civilization.[15]

As a true modern universalist, Planck specifically sought "units of length, mass, time and temperature which are independent of special bodies or substances, which necessarily retain their significance for all times and for all environments, terrestrial and human or otherwise."[16] As discussed in chapter 3, Planck was able to derive a new constant for his calculations from a fixed element size in relation to wavelength, called h. In the equation that would later become the basis for quantum theory as it was taken up by Einstein and other physicists, $E = h\nu$, the minimum quantum of energy denoted by h is a product of energy and change over time. It appears as a limit in calculating the very specific problem of black-body radiation. But as Planck came to realize the implications of his metalogical reasoning, he was inclined toward claiming the general significance of h as a universal constant. Considering h as nature's own fundamental limit, Planck was able to derive what are today called "base Planck units"—specific measures for length, time, mass, charge, and temperature. Physics today operates with five constants of nature, of which three are most general and applicable to cosmology. First, Newton's G for the gravitational constant. Second, *c* for Einstein's speed of light in a vacuum. And third, Planck's quantum of action, *h*. All three are involved in the equations from which Planck length, mass, and time are calculated.[17] This invention of hypological constants enabled an entire new order of mathematical relationships to be expressed. Logically, this could happen in one of two ways.

On the one hand, Einstein's idea of *c* is similar to Newton's *G* for gravity, as the hypological supposition that enables a mathematically coherent description of the universe. On the other hand, Planck's *h* emerges through metalogical calculation, or more precisely, as a quantitative expression for the limit condition of metalogical reasoning (statistics) within a hypologically constituted order (a black-body space). In turn, this logical difference bespeaks how physics today handles its limit conditions differently. The problem of gravity is solved hypologically by being integrated into the framework of general relativity. The problem of heat or energy, however, is solved metalogically through quantum mechanics. On one side, *G*—on the other, *h*. And the problem of light?

Insofar as light is considered exclusively as uniform speed—that is, as a wave—it is understood hypologically. But insofar as light is considered as a particle, it is understood metalogically. Thus, Einstein's invented constant *c* appears as the bridge between two logical orders of physics. The autological given we know as light is precisely where the logic of general relativity breaks down: if light is considered particulate, its photons must be without mass and therefore not subject to gravity. Light is the exception that confers stability on the system of general relativity as a whole. At the same time, light is the condition for quantum mechanics itself splitting into two orders of statistical constructs—bosons and fermions, or matter and interactions. Metaphysically, then, light is the fundamental condition for the bifurcated logical orders of explanation in physics.

What happens when the limit condition of light is mobilized against the limit condition of gravity, and applied across the entirety of a conceivable universal space-time horizon? This is one way to describe Einstein's general relativity theory, and the result is a level of complication that makes the theory exceedingly difficult to test empirically. As an example, consider that in Einstein's universalization of $E = mc^2$, the axiomatic relationship between mass and energy means that gravity curves space-time and thus also bends the path of light. But could gravity actually influence the speed of light itself? A photon, the "particle" of light, is technically considered to have zero resting mass and would therefore, according to general relativity, not be affected by gravity. But here the theoretical and experimental limits overlap in a self-reinforcing movement. Not only is the concept of resting mass itself hypothetical, but this specific kind of particle is by definition the limit of motion itself. Moreover, it has proved impossible to ascertain from experiments whether the photon, as it is understood by modern particle physics, actually has zero mass, let alone if such a value could ever be shown.

Ultimately, just like the experimental facts that constitute electrons and other elemental particles, the masslessness of light is not empirically verifiable but rather a necessary condition for upholding the framework itself. If light were not massless, then its speed could not be constant. The very exception of the photon from the physical distribution of mass among elementary particles is what guarantees the stability of the system within which it is given meaning. If light were to have even the slightest mass—and from an experimentalist position this cannot be ruled out—the entire framework of modern particle physics and cosmology, and the entire structure of unit standardization derived from it, would come unhinged. And the greater the scale, the greater the error of multiplication. In this sense, the constancy of the speed of light is not significant for its actual, calculable speed under hypothetical conditions but rather for expressing the limit condition of physics itself.

As the theory of general relativity garnered interest, Einstein suggested ways of testing its mathematical predictions. An observed discrepancy against the Newtonian theory in the axial rotation of Mercury by about 40 arcseconds (ca. 0.01 degrees) was the first indication. Einstein's general relativity could explain this as gravitation being mediated by the curvature of space-time and yielded calculations in closer agreement with observations. Immediately following World War I, general relativity had a sensational public breakthrough with an expedition by Lord Eddington that appeared to confirm Einstein's prediction about the bending of light, though the precision was poor and the results difficult to ascertain. For the three following decades,

Einstein's theory, despite his fame, attracted little to no interest, before it was revived as the centerpiece of an entirely new cosmology in the 1950s. What made the theory of general relativity so useful was not its accuracy but rather its scope—the ability to describe a universe far beyond our own solar system in one ostensibly coherent framework.

However, underneath Einstein's elegant construction of principles, making use of Planck's natural constants, which together appear to tie the cosmos together in a single mathematical formulation, there remains a problem of scale. There is a vast difference, quantitatively and qualitatively, between a theory that explains the dynamics of our own solar system (such as Newton's) and a theory that purports to account for the entirety of a universe of billions of light-years and galaxies. What has occurred in the historical development of the modern scientific universe is a kind of slippage from our galaxy to this ever-expanding totality, in which the conceptual coherence with the old system masks the exponential leap in explanatory ambition. For instance: What if the composition of galaxies outside ours differs from the conditions of our solar system? What if space-time constituents consist of variations that do not fit our calculations? It would not be difficult to imagine conditions that impose practical limits to the infinite reach of our theories. Already in 1922, in a popular account of the theory of relativity, the French mathematician Émile Borel put the matter in perspective:

> It may seem rash indeed to draw conclusions valid for the whole universe from what we can see from the small corner to which we are confined. Who knows that the whole visible universe is not like a drop of water at the surface of the earth? Inhabitants of that drop of water, as small relative to it as we are relative to the Milky Way, could not possibly imagine that beside the drop of water there might be a piece of iron or a living tissue, in which the properties of matter are entirely different.[18]

Borel's point speaks to the almost unimaginable scalar difference between what we know as the universe and our specific place in it as actors and observers—and he articulated this well before the universe was expanded far beyond our own solar system and explained in terms of its evolution in time. According to present calculations, based on an array of hypological constants, the currently estimated radius of the universe is 46 billion light-years, which roughly converts into terrestrial distance as 276,000,000,000,000,000,000,000 miles. A theory of the entire universe therefore requires a lot of simplifying assumptions, and when these assumptions are multiplied and stretched across vast distances, the potential for error increases.

Moreover, this grand mathematical world-picture relies on very limited means of testing. Besides the many scientific difficulties involved in inventing workable experiments, the technology for probing a galactic order of magnitude requires enormous capital investment. In the early post–World War II era, the fledgling field of cosmology did not enjoy the major capital influx that dominated the physical sciences in the 1940s and 1950s. Therefore, until cosmology could call itself Standard Cosmology, in concordance with the Standard Model of physics, the very few successful experimental observations that existed were highly significant in structuring the terms of discourse.

In fact, the midcentury period in which cosmology would be transformed into a scientific discipline can be defined as the interval between two key astronomical observations: the expansion of the universe, in 1929, and the cosmic microwave background radiation, in 1965. Neither case was strictly an observation as such, but each succeeded in being mediated as a scientific discovery within a comprehensible frame through a complex interplay of theoretical and experimental predictions based on massive machines and pliable parameters. And as we shall see, a certain hypological ingenuity to make it all come together.

HAWKING'S SINGULARITY

When Einstein worked out his complicated set of field equations for general relativity, he was drawing on a constellation of three principles—relativity, constancy, and equivalence. However, the theory still lacked a fourth principle by which it would be complete and bounded. In analogy to classical philosophical doctrines of causality, this could be called its teleological dimension—its boundary condition. To Einstein, general relativity was a field theory. But what kind of field?

How could it be circumscribed? The cosmological battles of the mid-twentieth century hinge on the rivaling answers to this question.

Whereas the classical Newtonian theory of the universe considered space an infinite void, Einstein rather favored a "space-bounded, or closed, universe," for which he provided three arguments. First, pragmatically, "from the standpoint of the theory of relativity, to postulate a closed universe is very much simpler than to postulate the corresponding boundary condition at infinity." Second, epistemologically, "it is more satisfying to have the mechanical properties of space completely determined by matter, and this is the case only in a closed universe." Third, probabilistically, "an infinite universe is possible only if the mean density of matter in the universe vanishes. Although such an assumption is logically possible, it is less probable than the assumption that there is a finite mean density of matter in the universe."[19] To the question of mean density, later denoted with the Greek letter Ω, omega, Einstein could only reason by assumption—but later in the century, as probabilistic reasoning became more entrenched, it would become of key importance.

Nonetheless, the elegance of making space determined by matter turned out to have a troubling consequence when calculations were performed. Due to the self-reinforcing quality of gravity through the distribution of matter, according to general relativity the space-time of the universe would have to expand at great distances. Because the prevailing understanding of the universe at the time was, as Arendt reminds us, a continual process without beginning or end, Einstein decided to counteract his undesirable mathematical answer by introducing what he called a cosmological constant, a negative pressure equal to the positive expansion predicted by the field equations. Einstein admitted that there was "no physical justification" for this constant and it would complicate the theory and thus reduce its logical simplicity—yet it appeared a necessary limitation. In this sense, the cosmological constant constitutes the fourth principle of the original theory of general relativity, in which the universe was guaranteed to be unchanging in its boundary state.

However, the modern view of the cosmos began to change by the end of the 1920s after Edwin Hubble published the findings from his telescopic research on the galaxies beyond our own solar system.[20] By identifying individual variable stars as "standard candles" for measurement, Hubble could establish a roughly linear relation between velocities and distances among nebulae for which velocities have been previously published—a linearity that would later become known as Hubble's law for all objects moving in an "empty" universe of idealized homogeneous properties. Hubble's measurements were based on the perceived redshift of the stars, the shift of light toward the less energetic part of the light spectrum as it moves away from an observer—a phenomenon known since the nineteenth century as the Doppler effect. The extent to which the Doppler effect would pertain to outer galaxy stars, as well as the extent to which outer galaxy stars can be used as "standard candles" for measurement, would be subject to much argument, observation, and theorization in the following decades. Within a few years, though, Hubble's general conclusion succeeded in mobilizing interest among astronomers. If his observations were correct, outer galaxies were moving away from us with a speed proportional to their distance. The farther away they were, the faster they appeared to be moving, at a linear rate denoted by the Hubble constant. The universe appeared to be expanding, and this in turn had consequences for Einstein's theory.

After Hubble's findings, a set of solutions to Einstein's relativity equations emerged, in which the cosmological constant was omitted and universal expansion made possible. These solutions were worked out independently, first by the Russian mathematician Alexander Friedmann and a few years later by the Belgian theoretical physicist Georges Lemaître. In place of the cosmological constant, Friedmann had to make a different fundamental assumption: the universe as a whole is spatially homogeneous and isotropic—that is, the same in all directions, from any perspective within it—but temporally variant.[21] Lemaître, who was also a Catholic priest, eagerly pursued the mathematical inference of cosmic expansion through time. In 1931 he published the first in a series of papers touting the "primeval atom" hypothesis, based on the invention of the particle, suggesting that "we could conceive the beginning of the universe in the form of a unique atom, the atomic weight of which is the total mass of the universe . . . [and which] would divide in smaller and smaller atoms by a kind of super-radioactive process."[22] Lemaître's boundary condition on general relativity, in other words, implied that the universe would be an evolutionary field of finite age arising from the properties of a nuclear particle formation. But throughout the 1930s the attempt at explaining creation through physics was resisted by much of the community, which was still dominated by astronomical observations, none of which supported Lemaître's highly speculative theory.

It took a nuclear physics research program in the United States from 1940 to 1953, enabled by the spike in federal science funding, to develop a theory of early universe cosmology that would draw the explicit link between an expanding universe and a finite origin of this process within the framework of general relativity. Russian-born George Gamow, without knowledge of Lemaître's earlier work, outlined a theory of an explosive nucleo-synthesis at the beginning of space-time. To be precise, the big bang theory, and all its variations since, only speaks to the immediate aftermath of an explicitly hypothetical event. A paper by Gamow's colleagues offered calculations for the emergence of elementary particles starting at .0001 seconds after "$t = 0$", corresponding to a temperature of 10 trillion degrees Kelvin, and for the following 600 seconds. What their work described was the limit condition of the universe, the point at which the theoretical framework breaks down. The theory says nothing about the event, or even if there could be one, but the working hypothesis to run a set of calculations soon became formulated as an absolute alpha point of nuclear expansion in the entire universe.

In a BBC radio broadcast in 1948, Gamow's conception was first christened the "big bang" theory by the British astronomer Fred Hoyle. He intended it ironically, as a derogatory term to distinguish Gamow's model from his own alternative cosmology, called steady state theory, but the irony was not widely appreciated. Instead, the name stuck and steady state theory quickly went on the defensive. As the name implies, steady state theory is aligned with the classical expectation of a cosmos unchanging in its boundary condition. The metaphysical premise for Hoyle and his colleagues was thus closer to Einstein's original sense of relativity theory as constituting a four-dimensional space-time. Instead of a cosmological constant, steady state theory introduced the "perfect cosmological principle," which holds that the universe is homogeneous and isotropic in all dimensions including time. That is, its apparent expansion in one place is supplemented by the continuous creation of matter at the heart of stellar constellations, a so-called C-field, which ensures universal isotropy overall. Like a modern statics theorem, what we see as expansion implies contraction elsewhere in order to keep the cosmos in equilibrium.

In Britain in particular, the steady state challenge spurred a public debate that brought out latent theological concerns. Hoyle attacked the big bang theory not only on its scientific terms but for its political, ethical, and religious consequences. Critical of organized religion, Hoyle suggested links between Christianity and big bang theory that he found preposterous and argued that steady state theory left no metaphysical room for Christian belief. To be sure, any lingering suspicion of big bang theory's consonance with a certain miraculous event was not quelled by Pope Pius XII, who declared that the theory was in perfect harmony with Christian belief and that modern cosmologists thus had arrived at the truth theologians had known for millennia. Meanwhile, on the other side of the iron curtain, Soviet communist leaders declared the big bang theory an ideological enemy of materialist science, because they thought the universe, following the prevailing view, must be infinite in both space and time. However little this mattered to the scientific merit of Hoyle's theory, it likely didn't do steady state theory any favors in the West to have his view opposed by the pope and acclaimed by the Kremlin.

Inside the scientific community, the ongoing discourse turned into a battle of competing metaphysical principles. Opponents of the steady state theory dismissed the conception of the C-field as a violation of the principle of energy conservation, whereas proponents argued that the perfect cosmological principle had to hold more absolute than a probabilistic law of thermodynamics. The dividing line between science and nonscience was made more tenuous by debates on whether cosmology as such was properly scientific. Steady state theorists repeatedly invoked the British philosopher of science Karl Popper to argue that their theory was more open to falsification through astronomical observation than the rival, and they were repeatedly countered with the claim that their cosmology was predicated on aesthetic preferences that lacked support in existing physical data. In both cases, the charge of overreaching empirical knowledge was leveled at the other side and "metaphysics" routinely hurled as a term of abuse. The cosmological contest of the 1950s played out as a great clash of positivist universalisms.

Eventually, the event that ostensibly tipped the scales in favor of the big bang was a strange, random event. Around 1965, two radio astronomers in New Jersey were conducting tests on a large microwave receiver designed for satellite communications. They inadvertently found that all their tests exhibited the same constant background noise in all directions. Having no idea what it could mean, they came in contact with a group of theoretical cosmologists at Princeton who happened to be seeking experimental support for their own prediction that, given the hypothetical big bang event, the universe today should be filled with feeble radiation. In a later critical review of the finding, the American cosmologist Geoffrey Burbidge argues that the ostensible conjunction of predictions and observational data in the discovery of the microwave background radiation was very much ad hoc. Examining the original calculations, he shows how it adopts a key numerical coefficient in order to make the calculated value agree with the observed value. "This is why the Big Bang theory cannot be claimed to explain the microwave background. . . . It is only an axiom of modern Big Bang cosmology, and the supposed explanation of the microwave background is a restatement of that axiom. Thus in no sense did the Big Bang theory predict the microwave background."[23]

However, once the connection was forged, there was a three-way retroactive reinforcement: experimental observation, mathematical projection, and metaphysical grounding converged on the big bang theory in much the same way as inferences of the expanding universe had coalesced with Einsteinian relativity in the late 1920s. This had to be proof. The new variable known as microwave background radiation was quickly seized by big bang theorists before steady state theorists could work out its implications, and it gave them an edge. As Kragh describes it, by the end of the 1960s the big bang theory, consisting of a large class of models sharing the assumption of a hot, dense beginning of the universe, had become a standard theory accepted by a large majority of cosmologists. In fact, it was only from this time that cosmology emerged as a scientific discipline and "cosmologist" appeared as a name for a professional practitioner of a science, on a par with terms such as "nuclear physicist" and "organic chemist." Although rival cosmologies did not disappear, they were marginalized. Not only was the Big Bang now taken to be a fact, rather than merely a hypothesis, it was also taken for granted that the structure and development of the universe were governed by Einstein's cosmological field equations of 1917.[24]

Perhaps the most significant difference between the big bang and steady state theories lay not in their respective scientific rationales but in their disciplinary grounding. The big bang theory grew out of nuclear physics and initially found scant support among astronomers. The steady state theory, on the other hand, showed strong correlation with prevailing astronomical observations, but its account of matter creation failed to convince physicists. As both theories in the 1950s and 1960s went through readjustments, the structural increase in research funding for nuclear physics played in favor of the big bang theory in the longer run. Above all, the big bang theory offered a lucrative and productive convergence with nuclear physics experiments, because the theory essentially made the universe itself into the ultimate particle accelerator. Under the regime of mass–energy equivalence, the higher the energy, the smaller the wavelength under observation, the deeper into nature at its smallest scales are we able to probe. And because big bang theory (unlike its rival) was built on a variant temporal dimension of general relativity, this meant that the higher the energy, the further into the early universe physicists could claim to see. Once the connection was forged, there was no way back.

In this sense, the particle accelerator signifies the pincer movement of nuclear physics and cosmology in the twentieth century, a mutual constitution whereby underground experiments can re-create the original conditions of the universe and astronomical observations can be used to determine the character of nuclear physics. Today, relativistic cosmologists study the skies just as nuclear physicists study particle collisions—looking for debris from which they can deduce the inner structure of matter according to their theories. In both cases, the assumption of an initial hypothetical event is the same, as is the standard set of parameters from which calculations can be made. The overlapping agendas of cosmology and particle physics became increasingly intertwined, and the disciplinary merger only strengthened the sway of big bang theory.

What happened to the steady state theory after the microwave event? Caught off guard by the sudden shift in the discipline, a few of Hoyle's followers began working on models that could account for the observed radiation phenomenon within the existing framework. Some were able to reproduce the same predicted results without inferring an evolving universe, including a later version called the quasi–steady state theory, which claims, based on a cyclical model of the universe, to better explain existing astronomical data. But whatever its merits, it was already too late. In the following decades, scientists and historians would continue to retroactively constitute and streamline the history of the big bang theory as a tale of inevitable discovery. The hypological origin had, through consistent and concerted action, become a foundational fact.

To this crucial consolidation phase belongs the notable work of the late British theoretical physicist Stephen Hawking, whose brash drive toward the theoretical unification of physics succeeded in capturing the imagination of scientists and nonscientists as the late-twentieth-century inheritor of Einstein. Hawking's distinctive brilliance resembles Einstein's in his ability to reinvent the limit condition of the theory as a foundational feature of the framework itself. This is how he used the mathematical concept of a singularity, a term for the point at which mathematical prediction breaks down. Metaphysically, a singularity functions as an exception to the very system within which it is given meaning.[25] In the Friedmann-Lemaître equations of general relativity from the 1920s, singularities had kept cropping up, something Einstein had profoundly disliked and which Lemaître had circumvented in his theory of the "primeval atom." Hawking, however, found a way to capitalize on this ostensible error by drawing on the work of the mathematician Roger Penrose, who first applied the concept of the singularity to the universe. According to Hawking, Penrose showed that

> a star collapsing under its own gravity is trapped in a region whose surface eventually shrinks to zero. . . . All the matter in the star will be compressed into a region of zero volume, so the density of matter and the curvature of spacetime become infinite. In other words, one has a singularity contained within a region of spacetime known as a black hole.[26]

Thus the singularity that appeared as an anomalous mathematical condition in the Friedmann-Lemaître equations was reinvented as a particular cosmological state. The metaphysical essence of the black hole argument is that light and gravity, both limit conditions of the framework itself, can be mathematically recombined to predict a physical phenomenon. In other words, if general relativity is correct, black holes must exist somewhere in the universe.

However, the problem with the singularity is that it cannot be represented. As Hawking puts it, "The singularities produced by gravitational collapse occur only in places, like black holes, where they are decently hidden from outside view by an event horizon."[27] Nobody had seen a black hole, because it could not be seen. But drawing on blackbody theory from quantum physics, Hawking was able to infer that a black hole would emit radiation and particles in a way that might make it detectable. Hawking's influential work sent cosmologists and astronomers off to verify the theory. Four decades later, several contenders for possible black holes have been suggested, but none has been empirically demonstrated. Nevertheless, the concept has made its way into the public imagination as well as that of scientists and appears to have completed its hypological circle to become a self-evident fact. The idea fit all too perfectly with the emergent cosmological model.

Hawking's next decisive move, in the spirit of rational mechanics, was to turn the temporal direction of the black hole postulate around. As he writes, "Penrose's theorem had shown that any collapsing star *must* end in a singularity; the time-reversed argument showed that any Friedmann-like expanding universe *must* have begun with a singularity."[28] In this way, Hawking was able to formulate how the framework of general relativity itself predicts the event of the big bang as an absolute alpha point. The equivalence of cause and effect is here turned into the invention of a new identity. As a singularity, the event of the big bang marks not only the limit condition of the universe where "the laws of science and our ability to predict the future would break down," but simultaneously the constitutive origin of the universe as such. Thus, a mathematical singularity suggesting a phenomenon called black holes (yet to be found) was reconstituted as physical fact to consolidate the framework from which it was initially derived.

In my analysis, the theoretical path from Einstein to Hawking constitutes the great hypological circle of general relativity: what began as a mathematical discrepancy was turned into a hypothesis, which was reinvented as a premise already inherent in the theory itself. It is so because it was made to be so. And in turn, the big bang became a self-evident and rather unquestionable fact to guide all future research and thinking about the cosmos. With the paradigm of cosmology properly enframed, it would enter a new phase of what Kuhn calls "puzzle-solving," putting the missing pieces together. But due to the vast complications of the total picture, as new pieces are found the puzzle keeps changing, creating new problems and new efforts to cram the pieces back inside the expanding frame.

RETROFITTING THE UNIVERSE

What is the universe? According to scientific cosmology, our solar system is located in one of many galaxies in the universe, distributed uniformly and the same in all directions. The universe has neither an edge nor a center, but it had a beginning, being created 13.8 billion years ago from a fixed amount of energy and matter, expanding and then cooling to allow the first atoms to form, and is still expanding at an increasing rate, with the majority of its mass being in an unknown form. Such is today's prevailing view. But how do we know this? And how do we know that we know?

Big bang cosmology is not a single theory but rather, as the astronomer Michael J. Disney puts it, "five separate theories constructed on top of one another."[29] The ground floor consists of Lemaître's expansion model based on general relativity, which connects Einstein's principles to Hawking's singularity theorem in light of a few specific observations, resulting in a complicated mathematical beast developed over roughly sixty years. Disney assesses this part of the theory, which laid the foundation for the professionalized discipline of cosmology in the 1970s, as "moderately well-supported" by the few observations that exist, but he argues that more direct evidence is needed to warrant the certainty with which this fundamental part of the theory is treated. As Lindley puts it, "General Relativity remains even today one of the least well tested of physical theories. It has passed all tests to which it has been put, and it is more elegant than any rival theory of gravity, but its predominant position in physics rests largely on its power to connect in a coherent and beautiful theoretical framework a handful of experimentally reliable facts."[30] Thus, the first floor of big bang theory rests on a set of fundamental principles and assumptions, including a model of space-time, that cannot itself be tested but whose main purpose is to simplify the conditions for workable mathematical solutions, provided in part by Einstein's field equations.

On top of expansion is a hypothetical "inflation" theory, which conjectures that the early universe went through a rapidly accelerating phase. When the inflation theory first appeared in the 1980s, proposed by the American cosmologist Alan Guth, it became a sensation and was eagerly supported by Hawking and other cosmologists because it allowed cosmologists to account for discrepancies left over from the expansion model. If Lemaître's model is the ground floor, inflation is the "pump" of the entire big bang construction, which can explain the missing link in the theory between how the hypothesized nucleus event could develop into conditions for the kind of galaxy formation that we can observe today.

In 2017 the inflation theory came under serious attack from senior researchers in the field, published in a *Scientific American* article that generated heated debate. Anna Iljas, Paul Steinhardt, and Abraham Loeb argued that the theory must be discarded because the only specific forms of inflation that could theoretically account for our own universe no longer appear to fit recent data from satellites and telescope observatories. Moreover, they argued, echoing the objection of

Disney and others, the theory has become so all-encompassing and complicated with so many pliable parameters that it can be tweaked and adjusted for any kind of observations, and one model of inflation can be replaced by another as new observations appear to contradict it. In their view, the theory is therefore no longer scientific. Their critique was met with an outraged response from Guth, Hawking, and thirty-one other physicists and cosmologists in a signed letter. The critics contend, wrote Guth and his colleagues, that

> inflation is untestable because its predictions can be changed by varying the shape of the inflationary energy density curve or the initial conditions. But the testability of a theory in no way requires that all its predictions be independent of the choice of parameters. If such parameter independence were required, then we would also have to question the status of the Standard Model, with its empirically determined particle content and 19 or more empirically determined parameters.[31]

The critique and its response, in other words, concerns the limits of what is scientific, and what drew the particular outrage of Hawking and his peers was the suggestion that the current theories of cosmology operate outside the bounds of empirical science. The goalposts of what is considered scientific and what is not have moved considerably in the course of the twentieth century, with the invention of new methods in theoretical physics being used to defend similar boundary-pushing in other disciplines. Where the line of empiricism ends is itself a metaphysical question, and scientists practicing at the very limit of science will of course defend their methods of mathematical conjecture, just as surely as scientists with alternative theories will attack the prevailing models on the same basis. The flash debate succeeded in showing how necessary something like inflation is to the coherence of the framework of cosmology, and it is perhaps not surprising that cosmologists will prefer to tweak individual parameters rather than question the structure of the edifice itself. With the wiggle room that the current model allows, it would take more than a few vocal critics to discard a theory that engages so many scientists.

Along with the introduction of inflation into the big bang theory, a "third floor" was also added in the course of the 1980s, the so-called dark matter hypothesis, which explains observations of galaxies that seem to contradict the expansion model through the postulation of an unknown form of matter. In fact, the theory requires nearly 27 percent of all mass-energy content in the universe to be dark matter, whose only reason for existence is to make the cosmological field equations add up. "Most cosmologists welcomed the possibility of such dark matter, because it might be lumpy enough to get the galaxies formed in the early universe—another serious problem for theorists," writes Disney.[32] According to calculations, assuming general relativity holds true in a far distant galaxy from our own, dark matter dominates and only about 5 percent of the universe is made up of "ordinary matter" that we can observe in faraway galaxies. A new, fourth floor of the theory is occupied by "cold dark matter," another hypothesized form of invisible constituent. Cold dark matter would not react to radiation and therefore could explain later formations that dark matter could not. By allowing for these mysterious and unseen elements to exist in huge quantities and behave in very specific ways, the overall theory could still account for current observations without falling apart.

Then, in 1998, a set of supernova observations of spiral galaxies concluded that these strange objects in the sky appeared to be spinning too rapidly to be held together by the mutual gravitational pull of their observable contents. Presuming the theoretical framework is correct, astronomers concluded that something else beside visible and invisible matter must be present that could account for the coherence of these galaxies. Thus another floor of the big bang theory was created for the mysterious force that must be responsible for this acceleration. According to present calculations, dark energy would have to account for almost 70 percent of the overall mass–energy of the universe. Ironically, this has led to the reinstatement of a cosmological constant that Einstein first wrote in as a fictional necessity, and which was later withdrawn. On the fifth floor, the cosmological constant was renamed dark energy and got a penthouse floor (for now), a necessary addition to the building to account for the widening gaps between theory and observation. Within each of the floors below, there is continual retrofitting and remodeling as new variables come into play, making the big bang theory a dynamic venture of puzzle-solving in the eyes of its proponents. Or is it rather a house of cards?

Much like the seven unknown dimensions of M-theory discussed in chapter 1, dark matter and dark energy are names for that which is required for the mathematical coherence of the theory, most fundamentally determined by the metaphysical requirement of unification: simply put, that one and the same mathematical framework invented on Earth must be valid for all of galactic space. Disney's critical review of the theory concludes that the current concordance model of cosmology consists of seventeen independent parameters, whereby only thirteen can be fitted reasonably well to the observational data. "Cosmology has always had such a *negative significance*, in the sense that it has always had fewer observations than free parameters, though cosmologists are strangely reluctant to admit it." He concludes: "A skeptic is entitled to feel that a negative significance, after so much time, effort and trimming, is nothing more than one would expect of a folktale constantly re-edited to fit inconvenient new observations. . . . While it is true that we presently have no alternative to the Big Bang in sight, that is no reason to accept it. Thus it was that witchcraft took hold."[33]

Paradoxically, the increasing complication of observable and experimental phenomena appears to have only intensified the faith in a theoretical framework within which the entirety of celestial phenomena can be represented. As in the case of string theory, the thrust of explanation has been inversed from Einstein's time: cosmologists today, whether they sift through data from a particle accelerator or from a space observatory, are searching for events that might fit their hypothetical framework, rather than the other way around. The development of big bang theory has turned Earth itself into a special case of a mathematical universe. While the physics of big bang cosmology become more internally unstable and precarious, its actors work steadfastly toward the goal of mathematical unification.

Aiding in the construction of this theory is a set of "derived Planck units," extending calculations from Planck's constant h to the base units for all established dimensions of physics, from area and volume to momentum and impedance. Quantitatively miniscule, in the negative exponential range of 30 to 40, they are mostly used by cosmologists who try to constitute the limits of the universe. The so-called Planck barrier, made up of these limiting units, constitutes the logical constraint of the scientific universe, against which it is continually explicated. It is the parameter for how to understand the big bang event, black holes, white dwarfs, and all other singularities, and they function as the mathematical boundary states of a calculable universe. In turn, these constants have undergone their own reunification in the form of the so-called fine structure constant—an abstract constant of constants. Dividing the square of the quantum electron charge by the multiplication of the Coulomb constant with the Planck constant and the speed of light in a vacuum, the fine-structure constant is known as a dimensionless coupling constant—dimensionless in the sense that it will by definition carry the same numerical value in all systems of units. Considered a "pure number," untarnished by any material phenomenon, the fine structure constant is thought by some to operate as a kind of meta-parameter for the universe and by others as the mathematical expression of nature itself. The reality behind appearances, in this view, is mathematics. As Ian Hacking puts it:

> Many cosmologists of today entertain the following picture. The universe is constituted first of all by certain deep equations, the basic laws of everything. They are composed of variables for measurable quantities, and free parameters whose values are fixed by assigning constants. . . . Then various boundary conditions are added, conditions not determined by the equations and the fundamental constants. . . . Such a cosmology is not far removed from Galileo's theism and his picture of God writing the Book of Nature. The Author of Nature writes down the equations, then fixes the fundamental constants, and finally chooses a series of boundary conditions.[34]

Planck succinctly summarized the operative mode of the mathematical universalism that connects Galilean invariance to big bang theory: "The increasing distance of the physical world picture from the world of the senses means nothing but a progressive approach to the real world."[35] A "progressive approach to the real world" is predicated on the very disappearance of reality. In cosmology, as in physics, theoretical laws are fundamental, but phenomenological laws are circumscribed and overturned. And to accomplish such a radical inversion requires, as Planck knew well, a deeply rooted belief: "Over the entrance to the gates of science's temple are written the words: Ye must have faith."[36] But a faith in what?

THE UNIVERSE OF HUMANISM

In 1980, Hawking delivered a lecture in which he suggested that the reunification of physics was in sight and that new mega-experiments (such as the JSWT and the LHC) would likely complete the major pieces of the physics puzzle. Afterward he toned down his claim but nevertheless made it clear that the stakes of reunifying the theoretical bounds of the universe are far higher than securing the operational ease of physicists. For Hawking, the question of the origin of the universe is explicitly theological in character. In the final paragraph of his 1988 best seller, *A Brief History of Time*, he writes that if physics discovers a complete theory of the universe, it would be a crucial step to "the question of why it is that we and the universe exist. If we find the answer to that, it would be the ultimate triumph of human reason—for then we would know the mind of God."[37]

Hawking's messianic cosmology at the tail end of the twentieth century leads directly to the question, as he puts it, of why it is that we and the universe exist. Most significantly, what does it mean that we as beings have learned the identity of the universe and its cosmic truth? Scientists are fond of the "objective" Copernican perspective that we are not the privileged center of the universe, but nevertheless it is we who are doing the observing and theorizing. And so Hawking's question essentially concerns the classical problem of the relation of humans to the universe, "man and nature." Now as then, God lurks in the background.

The essential symmetry between the universal and the human, inherent to modern metaphysics from its seventeenth-century constitution, reaches its apotheotic expression in what is known as the anthropic principle. Coined by the Australian physicist Brandon Carter in the 1970s, the anthropic principle says that, as Hawking puts it, "we see the universe the way it is because we exist."[38] To physicists, the anthropic principle prescribes how one is to account for the relation between our particular situation as observers in a universe and the universe as a whole when considering small data samples within it. Its explicit function is as a probabilistic weighting of the bias inherent in any observation. Carter argues that the anthropic principle emerged in response to two existing contenders for describing the human–universe relation. On the one hand is the "pre-Copernican dogma" of the autocentric principle, which places human terrestrial observers at the center of the universe. On the other hand is the perfect cosmological principle (employed by steady state theory), which holds that the universe has no privileged center, is homogeneous and isotropic, and that our local area can therefore be considered a typical random sample. For big bang theory, neither of these principles would work, because while our planet is clearly not at the center of the general relativity universe, it is nonetheless in a very particular stage of an evolutionary process that distinguishes it temporally from other parts of the universe. This leads to the postulation of the anthropic principle, which Carter presents as "a reasonable compromise between these unsatisfactory over-simplistic extremes" (conveniently glossing over the fact that the extreme oversimplification is due to his own rhetorical construction). Carter explains:

> The anthropic principle would have it that . . . the a priori probability distribution for our own situation should be prescribed by an anthropic weighting, meaning that it should be uniformly distributed, not over space-time . . . but over all observers sufficiently comparable to ourselves to be qualifiable as anthropic.[39]

He contends that the anthropic qualification is not simply an identification of the human, nor a restatement of Earth-biased autocentrism, because the principle is intended to encompass "extraterrestrial beings with comparable intellectual capabilities."[40] Moreover, Carter argues that because it involves probability distribution and not absolute values, the anthropic principle is not actually tautological. Logically speaking, this is correct, as the anthropic principle is rather hypological in character. It does not validate itself directly but rather works to circumscribe the framework within which it was first constituted—it gives shape to the questions and conclusions drawn from the limited field of cosmo-logical observations. Whatever its possible practical uses to physicists, the anthropic principle has immense metaphysical implications, since it effectively completes the hypological circle between "human" and "nature" within the framework of mathematical universality.

Hawking reasons that the anthropic principle explains why the big bang occurred so many million years ago, because this is the timescale it would take for "intelligent beings" (presumably like cosmologists) to evolve. This is how he describes the process of universal formation:

> An early generation of stars first had to form. These stars convert-ed some of the original hydrogen and helium into elements like carbon and oxygen, out of which we are made. The stars then ex-ploded as supernovas, and their debris went to form other stars and planets, among them those of our solar system, which is about five thousand million years old. The first one or two thousand million years of the earth's existence were too hot for the development of anything complicated. The remaining three thousand million years or so have been taken up by the slow process of biological evolu-tion, which has led from the simplest organisms to beings who are capable of measuring time back to the Big Bang.[41]

In other words, within one and the same theoretical framework, the mathematical derivation of a singular event origin is given physical meaning within the context of a universal history, and this universal history is in turn justified in terms of the beings who invented it—a forward and backward movement of the same reinforcing logic. We, at this moment in history, have reached an intelligence advanced enough to explain our own cosmogony, and the story of this cosmogony is verified by our existence at this moment in history. Thus we appear to have made our own history transparent to ourselves in a complete framework within which our only remaining task is to fit a few pieces of a predetermined puzzle. As Heidegger might put it, our enframing by this metaphysical world-picture is total.

Nevertheless, the grand irony of scientific cosmology is that it suc-ceeds in reaching mathematical universality only at the cost of leaving our own world behind. It turns our own planet into an exceptional case of a metalogically and hypologically constituted universe and turns "humanity" as the "discoverer" of this mathematical universe into an exceptional condition for there to be a cosmos in the first place. Although the formulation of the anthropic principle theoretically opens to the presence of other beings, these beings are already defined in terms of having "comparable intellectual capabilities" to ourselves in order to qualify for anthropic status. If they are not us they have to be just like us by the implicit standard of scientific intelligence. Humanity, for which this scientific achievement would be the ultimate marker, is therefore the mirror singularity of the cosmos, that which gives the cosmos not merely its meaning but also its purpose. As Hawking puts it with deceptive simplicity: "We see the universe the way it is because we exist." Or put differently, we are ourselves the apex of cosmic evolution.

This conviction of cosmologists, physicists, and other practitioners of modern science is first and foremost a faith in a kind of progress that ascribes truth to current developments and myth to the theories of the past. Most conventional histories of science therefore contain variations on the formula, "Once upon a time people *believed* that [insert laughable myth here], but now we *know* that [insert selected experiment with supporting theory here]." In this rhetorical frame, the prevailing theory is of course always closer to the truth. But might this mode of progress be the most enduring myth of all? In spite of the many advanced forms of calculation and our near-infinite galactic probing with limitless data sets, is our understanding of the cosmos today really so much more profound than that of our ancestors?

Whereas quantification yields precision, understanding emerges qualitatively from seeing the matter with the right frame of mind. A scientist today would be humored by the Aristotelian divide between terrestrial and celestial physics as something modern knowledge managed to surpass—and yet physics is still split between two phenomenal scales reconcilable only through mathematical inventions such as an eleven-dimensional string-based universe. A cosmologist today might laugh at Descartes's antiquated vortex theory of the universe, even though all our telescopic images of distant galaxies reveal nothing but vortex formations in the exact image that Descartes was able to describe of his own accord. Physicists might chuckle at the ether theory from the late nineteenth century, when their peers "believed" that something in the universe must be mediating known phenomena, but today we "know" that the universe at the quantum scale consists of fields, which is of course a very different idea. Thanks to a mathematical theory, we now "know" that the universe consists of dark energy, black holes, and cold dark matter—and stretching this knowledge further, by recourse to "pure" numbers we may even predict the fate of the cosmos and postulate a multiverse and seven hidden dimensions of reality. The only reason to think such things exist is because they would be required by an unquestioned metaphysical imperative to explain all physical phenomena by one and the same theory. Read in a historical and metaphysical context, the laughter of today's advanced knowers rings hollow.

Through the scientific claims to knowledge, which may individually be refuted, replaced, and qualified, runs a more persistent claim of mathematical access to a higher universal truth, a claim that is not so readily allowed for the earthly sciences. Certainly, scientific cosmology has come a long way since Galileo, but four hundred years later, might an observer at the margin of the discipline be forgiven for wondering if big bang theory is not really Christian metaphysics by different means? While the big bang theory is by itself no religious doctrine (the pope's enthusiasm notwithstanding), it is the most authoritative metaphysical world-picture of our cosmos, residing squarely at the vanishing point between believing and knowing, a grand celebration of modern humanism in the name of mathematical universality, both mutually constituted dimensions guaranteed by an implicit, transcendental idea of God. As I have tried to show in this book, some of the fundamental metaphysical problems bedeviling modern physics and cosmology stem from this idea that still ties the science to its Christian origins: the unwavering faith in a *mathesis universalis*.

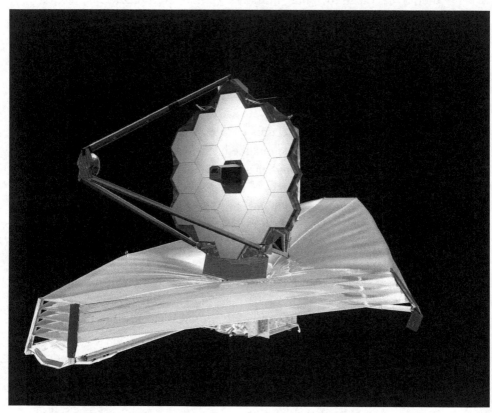

NEXT WORLD-OBJECT: a model of the James Webb Space Telescope. Copyright NASA, Creative Commons.

CONCLUSION
Lost in the Universe

It is scientists who ask the questions, and complexity arises when they have to accept that the manner in which they pose their questions has itself become problematic.

—*Isabelle Stengers*

Today, most of us have to travel far to be able to see it with our very own eyes.

Out of the cities, up on the mountains, or on transoceanic ships in the dark and clear night, far away from the polluting clouds of twinkling energy, radiating like hazy afterglow of the giant energy revolution that physics first unleashed, out where only the lights of the cosmos can reach us. Once it was immediately accessible to the naked eye on a clear night, but now the overwhelming abundance of stellar phenomena arrives predigested through internet pages, graphics apps, library files, maps and convenient factoids, measures, providing us with the simulacrum of understanding of the oceanic unknown that surrounds our stormy, lonely planet.

What Hannah Arendt called world-alienation, the perspectival shift she associated with Galileo's pioneering use of the telescope, contains an inverse dimension that has become more readily apparent today: our own alienation from the cosmos. Perhaps it is more than an irony of history that at the same moment in time as we are able to launch probing spacecrafts into the distant universe, the cosmos has largely disappeared from our direct view. Thus, we only increase our reliance on how the universe has come to be constructed for us.

The idea of the universe can of course be understood in many different ways beyond the framework of cosmology. What the scientific view lacks can be filled in with a range of alternative or spiritual ideas about the universe and its power, ideas that fall far away from the scientific purview but may still seem more significant to earthlings who want to relate to their cosmos more than they want to measure it. Nevertheless, the predominant scientific theory, even in its popular form, is the hegemonic framework for how we encounter the cosmos today—whether through the quantifiably accessible stuff of astrophysics or science fiction. Disappearing along with the fading lights in the sky from the horizon of history is the sense of everyday wonder that could accompany gazing at the nightly realms above, in which we could encounter at once ourselves and our own limitations. To look the cosmos straight in the eye, to sense its overwhelming power, is a moment of coming to terms with the autological dimension of the world, the force of an active presence, not an empty mathematical space but a fluid, magnetic, electric connection. Cosmic alienation in this sense means that the more scientific our understanding of the universe is, the less complex our relationship to it becomes.

When Galileo invented an experiment that restructured the idea of reality based on a fictional void, he performed a clever trick that allowed him to calculate motion on a large enough scale that the inevitable discrepancies between the actual world and his fictional model could be handled and were a small price to pay for the grasp it provided. When Newton elaborated on this maneuver and was able to submit our entire solar system to a set of simple laws that could bear a growing number of observations and calculations to a remarkable level of precision, it was by all means a towering achievement. Yes, it was based on a metaphysical trick, and the trick came with hidden implications that led to problems, but it worked to produce a very convincing map of our cosmos. The history of modern physics is an impressive register of such ingenious inventions and curious conceptions that have conquered mysteries through efficient explanations, and it has conferred on the discipline considerable scientific authority.

However, when Einstein's framework of general relativity was applied to the boundary conditions of the universe and the scale grew in the order of billions by billions times greater magnitude, so did the potential errors of multiplication. And somewhere along the way, the map began to determine the territory. No matter how advanced and precise the technology, the sheer scale of the contemporary cosmological endeavor stretches the limits of the credible. One problem is the limited set of empirical observations that makes the inferences so unconvincing. Another problem is that the framework provides no discernible way to assess its own validity.

In the great industrial metastasis of the twentieth century, modern cosmology became an entirely new kind of science with only a nominal coherence to the modern physics from which it emerged. And whereas the Standard Model of particle physics may have become vindicated by experiments at the LHC, there is no reason to assume Standard Cosmology holds similar sway over the galaxies. The conjoining perspectives between the micro and macro levels of physics emerge out of resource pragmatics, as experimental facilities to test theories at this level are in short supply, and more importantly, out of a metaphysical imperative toward theoretical unity for which there exists no physical justification.

Today, the universe in the scope of scientific cosmology appears as a hollow hypological construction, a name for something constituted by a multitude of metalogical parameters that may in and of themselves make sense of concrete and delimited phenomena but which never add up to a totality: a hypological universe that provides the semblance of unity from a reality too messy for mathematics as well as metaphysics. A simulacrum of unity, to what end?

Far beyond the ingenious procedures of a Maxwell or Einstein, today's science is not merely made from curious truth-seekers—its practice is inextricably tied to a technological order of giant machines, mega-industrial buildings, supercomputers, rocket launches, and explorer satellites. Scientists themselves are supposed to be disinterested, but for any institution, a principal interest lies in perpetuating itself. As Isabelle Stengers puts it in *Cosmopolitics*, "physics is confronted with the same kind of difficulty faced by every mega-enterprise threatened by bureaucratization and autism." Indeed, I think she is correct to observe that, contrary to the idealized self-conception of scientists inherited from a century ago, "a plausible future is within sight in which there will obviously be scientists, but they, as more or less competent employees, will no longer be distinguished from anyone else who sells their labor power."[1] Following this perspective, the purpose of the JWST, notwithstanding the individual motivations of the scientists within the enterprise, is to produce research, to engage scientific labor and create conditions for more capital-intensive projects in return.

Metaphysically, this science unfolds in many ways simultaneously, and in this book, I have suggested four "logical" dimensions of this activity. Scientists will carefully select data that allow them to leap to certain inferences and conclusions, analogically. Their results will have to mediate interest from other scientists, autologically. Operating in large networks, they will gather evidence from proliferating tests and data sets, metalogically. And those who succeed get their work recognized as bedrocks, building blocks, as events that can make history, hypologically. As Bergson realized more than a century ago, the objective of the scientific mind is not truth but action—what is at stake is how the production of new truths allows for a new reach, expansion, and mobilization.

The essence of technology, Heidegger wrote, is the logic of the framework itself, of the world-picture as the driving force of further mobilization of the physical world. What we call nature is progressively enframed, transformed, and left behind. We lack concise logical terms for the large-scale escalation and impetus to growth against all constraints, this leap out of earthly embodiment. The idea of the metalogical, as I have described it in this book, speaks to this rather untheorized dimension of the modern sciences and philosophy. As the classical conceptions of modern logic produce for us a world fundamentally conceived in terms of identity—as the universal and the particular, object and subject, nature and human—a gaping conceptual chasm appears between the autological striving of our individual existence and the totality of the world, besieged by multiplying forces beyond comprehension. Seven billion people, trillions of dollars, particles of 10^{-35} meters, galaxies billions of light-years away—the world as we know it is ever more difficult to conceive with an individual mind, and ever more reliant on structures, operations, and algorithms that reduce and simplify things within our grasp. Against these overwhelming metalogical vectors, it is perhaps not surprising that we tend to seek recourse to the most entrenched, axiomatic hypological conceptions that offer themselves innocuously as givens, simple explanatory structures to cling to like buoys in the swirling vortex of our world. This tendency appears as much in the sciences as in politics and culture, and in my view, the universe of scientific cosmology has become just such a conception.

After all, the image of the new space telescope, the JWST, beckons belief—in ourselves as humans, our power and potency. That we, "standing on the shoulders of giants," can build and launch a machine into outer space that reveals deep truths about our cosmos is a more appealing idea than its ostensibly dystopian opposite: that we, led astray by path-dependent war technology and the poorly understood implications of our past inventions, waste billions of dollars on machines that chase absurdities and sputter out vast databanks of hard scientific nonsense in an effort to drive technological innovation and increase economic output. It is perhaps harder to believe that such colossal effort could be mistaken, and easier to believe that the collective ingenuity that characterizes so much of human life on Earth also extends to the outer reaches of the galaxies. For who is to say what is true, if not the scientists themselves?

If my critique of modern cosmology appears to promote such a dystopian view, or be seen as a respectless attack on great human minds and institutions, or for that matter as a dangerous inquiry that could play into the hands of antiscientific worldviews, this is far from my intention. My years of studying the historical foundations of contemporary physics has led me to question the validity, credibility, and authority of many firmly held convictions in this science, but I can offer no alternative hypothesis or method that might serve the science better. Critique is not judgment, and I am in no position to judge. As Stengers reminds us:

> The manner in which the sciences and technosciences are presented today cannot be judged either as veridical or as false or as ideological, for it cannot be judged on the basis of an identity that would predate them. The manner in which they are presented is a part of their identity, like the production of all relations.[2]

In other words, despite all the reasons to be skeptical of the cosmological world-picture, and despite all the indications its paradigm may be creaking in its foundations, I can not claim that it is wrong, only that it is dubious. And I argue for a deeper questioning about the foundation of our sciences, open-eyed about the broader political context of power in which both the sciences and their critics operate.

In the end, it comes down to a question of belief. When the JWST eventually launches, or a similar astronomical mega-experiment takes place in the future, can we believe what it purports to tell us about our cosmos, through its computation and interpretation by cosmologists? My answer is that one might very well believe the science, and the science might be right. What I argue, based on history and philosophy, is that the only conceptual thread that connects the current world-picture of cosmology to the order of universal scientific validity is a transcendental structure of faith. Big bang theory can be believed to the extent that mathematical universality can be believed, for it is a fundamental, tacit metaphysical premise of the theory that no independent mathematical model or astrophysical experiment will ever validate or falsify. As Spinoza put it, he who has a true idea at the same time knows he has a true idea, and cannot doubt the truth of the thing. This is not to say scientific knowledge is equivalent to any kind of belief, only that our current world-picture is powered by this persistent historical form of faith. Perhaps the greatest problem for a science with such a metaphysical orientation lies not in the lack of empirical justification but rather in a blindness to its own lack of complexity.

Borrowing a distinction from Latour and Stengers, I understand complexity not in contrast to the simple but rather in distinction from the complicated. As Latour writes, complication "deals with series of simple steps (a computer working with 0 and 1 is an example); the other, complexity, deals with the simultaneous irruption of many variables. . . . Contemporary societies may be more complicated but less complex than older ones."[3] Complexity, in this sense, emerges autologically: it is tied to its own conditions of existence, whereas complication results from an order that radically separates an idea from its actuality.

The problem, as Stengers puts it, relates directly to the practice of scientific work and the question of *self-implication*, when the manner in which scientists pose their questions itself becomes problematic.[4]

Complexity, in other words, emerges from open engagement with the metaphysical limitations of a problem, and this perspective is what I find remarkably lacking in cosmology. Theoretical physicists are capable of dealing with problems to an almost infinite degree of complication, from eleven-dimensional string theories to redoubling statistical orders of bosons and fermions to black holes and white dwarfs in relativistic space-time—indeed, complication is what defines this discourse. But insofar as cosmologists and physicists continue to understand the universe as a structure independent of their own questioning, as a mathematical realm outside the history of its invention, they are emphatically not dealing with problems of complexity.

For Stengers, the formulation of complexity in distinction from complication allows for "the uncoupling of two dimensions that are often inextricably associated in discourses for or against the sciences" and together constitute the predominant image of scientific truth. On the one hand, the hypological constitution of scientific inquiry posits "the power of the analytical approach and the peremptory judgments that it appears to authorize." On the other hand, what I call the metalogical thrust of the sciences works to reinvent an overarching identity in place of a multitude of inquiries, reconstituting "a 'scientific rationality'" through the "production of 'scientific views of the world.'"[5] In the reciprocal capture of modern scientific discourse, the power and success of an analytical approach to any phenomenon appears to justify something like a "scientific rationality," and in turn, such a scientific rationality is precisely what warrants the "analytical approach" that appeared to make it so successful. By untangling these two dimensions, Stengers shows how the notion of complexity rather involves what I call an autological dimension, a mediation of scientists themselves in scientific problems.

Through an understanding of complexity, Stengers shows, the idea of truth as the goal of scientific inquiry is instead turned into a criterion of relevance. In this sense, we could say that Planck's natural constants, for instance, are not significant because they are universally true or unbiased by human activity (as Planck himself believed), but rather because their ostensible endurance as mathematical constraints makes them relevant to scientific practice. For a science capable of involving an autological dimension in its work, the problem is not to discover truth but rather to determine what matters and what does not.

> What is noteworthy about "relevance" is that it designates a relational problem. One speaks of a relevant question when it stops thought from turning in circles and concentrates the attention on the singularity of an object or situation. Although relevance is central to the effective practices of the experimental sciences, in their public version it often boils down to objective truth or arbitrary decision: to objective truth when the question is justified by the object in itself, and to arbitrary decision when it refers to the use of an instrument of experimental apparatus whose choice is not otherwise commented on. In the first case, the response appears to be dictated by reality. In the second, it appears to be imposed by the all-powerful categories of which the investigative instrument is bearer. Relevance designates, on the contrary, a subject that is neither absent nor all-powerful.[6]

Neither absent nor all-powerful means that cosmology and physics, and all the sciences that create our new givens of tomorrow, do not make themselves according to criteria that escape history. The question Stengers poses is how to reorient scientific inquiry toward the complexity of our world and simultaneously make it relevant to actors other than the ones already invested in the perpetuation of their inquiry.

As the JWST world-object is set to launch with great ambitions and risks riding on its back, and as some of us will make our way to a shoreline away from city light pollution to observe its meteoric rise into the heavens, the right question is not whether its metaphysical enframing of the universe is true or false—whether its explanatory power is absent or all-powerful. Rather, the question is whether, and to whom, the experiment is relevant. Will its eventual findings actually matter to anyone but the scientists themselves? Will the experiment matter to the nonscientific public as more than a noisy spectacle inspiring a fleeting sense of humanist awe in our capabilities?

Perhaps the space telescope will be a failure, or perhaps it will work exactly according to plan. Maybe it will find its place and manage to locate astronomical events that are taken as further validation of the framework that determined its design, while simultaneously indicating promising new avenues of research. Perhaps in five years of operation it will generate a decade's worth of science labor hours, dazzle the web with new imagery, and generate Nobel Prizes. But even in its most privileged view of the universe, it may just be as lost as the rest of us.

NOTES

Introduction

1. Serres, "Revisiting the Natural Contract."
2. Billings, "The Telescope That Ate Astronomy," 1030.
3. Billings, "The Telescope That Ate Astronomy," 1028.
4. Disney, "Cosmology."
5. Cartwright, *How the Laws of Physics Lie*, 13.
6. Cartwright, *How the Laws of Physics Lie*, 57–58.
7. Cartwright, *How the Laws of Physics Lie*, 72.
8. Stengers, *Cosmopolitics I*, 38.
9. Stengers, *Power and Invention*, 40.
10. Wolfe, *What Is Posthumanism?* xiv–xv.

1. Cosmology in the Cave

1. Heidegger, *The Question concerning Technology*, 115.
2. Heidegger, *The Question concerning Technology*, 56.
3. Latour, *Politics of Nature*, 10–11.
4. Latour, *Politics of Nature*, 34.
5. Stengers, *Cosmopolitics I*, 6–7.
6. Heidegger, *The Question concerning Technology*, 69.
7. Heidegger, *The Question concerning Technology*, 124.
8. Hacking, *Representing and Intervening*, 194.
9. Hacking, *Representing and Intervening*, 208.
10. Generally, I have in mind here the "revolt against metaphysics" associated with A. J. Ayers, as well as significant contributions by Moritz Schlick and

Rudolf Carnap, among others. For an account of the "event" at Davos in 1929 that could be seen as the splitting point of analytic and Continental philosophy (involving Carnap and Heidegger), also from a neo-Kantian perspective, see Friedman, *A Parting of the Ways.*

11. Hacking, *Representing and Intervening*, xii.

12. Hacking, *Representing and Intervening*, xiii.

13. Falkenburg, *Particle Metaphysics*, 209.

14. Falkenburg, *Particle Metaphysics*, xii.

15. Falkenburg, *Particle Metaphysics*, 3.

16. Stengers, *The Invention of Modern Science*, 90.

17. Stengers, *The Invention of Modern Science*, 80–81.

18. Stengers, *The Invention of Modern Science*, 89.

19. Here, Stengers references Latour, who develops the metaphysics of Serres. For a vivid account of the "parasitic" logic of mediation, see Serres, *The Parasite.*

20. Heisenberg, *Physics and Philosophy*, 63.

21. On this point in particular, see Bal, "Fundamental Forces." I will elaborate this discussion in chapter 4 as the central problem of general relativity.

22. Descartes, *Philosophical Writings*, 19.

23. Stengers, *Power and Invention*, 22.

24. In another prominent version of string theory, the idea of the universe is replaced by the multiverse, a cosmos consisting of different metaphysical realities. But even a multiverse implies mathematical universality in the sense that I will discuss in chapter 2.

25. Kragh, *Conceptions of Cosmos*, 222.

26. Lindley, *End of Physics*, 206.

27. Heidegger, *Identity and Difference*, 34–35.

28. Heidegger, *Basic Writings*, 277–78.

29. Heidegger, The *Question concerning Technology*, 173.

2. Of God and Nature

1. Arendt, *The Human Condition*, 259–62.

2. Arendt, *The Human Condition*, 251.

3. Descartes's correspondence is quoted from Gaukroger, *Descartes: An Intellectual Biography*, 219.

4. Gaukroger, *Descartes: An Intellectual Biography*, 290–91.

5. The titles of two of Galileo's works are translated with the term *dialogue*: *Dialogue concerning the Two Chief World Systems*, the 1632 publication that had him face papal inquisition, and *Dialogues concerning Two New Sciences*, the

clandestine 1638 text. Following classical convention, I will refer to the former as *Dialogo* and the latter, my key Galilean text, as *Discorsi*. I will use similar shorthand for Descartes and Spinoza.

6. Heidegger, *Identity and Difference*, 25.

7. Heidegger, *Identity and Difference*, 25–26.

8. Heidegger, *Identity and Difference*, 29–31.

9. This quote is from Heidegger's close reading of Leibniz's correspondence in *The Principle of Reason*, 119. Leibniz outlines the difference between the two principles several places, most succinctly in the *Monadology*, principles 31–32.

10. Serres, *The Birth of Physics*, 21.

11. Gaukroger, *Emergence of a Scientific Culture*, 23.

12. Newton, *Principia*, lxvii.

13. Gaukroger, *Emergence of a Scientific Culture*, 413. The relation between mechanics and mathematization has been part of Gaukroger's research since *Explanatory Structures* (1978), as well as his many books on Descartes.

14. See Biagioli's interesting account of Galileo's self-fashioning in *Galileo, Courtier*, esp. 6–17, 59, and 357.

15. Galileo describes the plane itself as "a piece of wooden moulding or scantling, about 12 cubits long, half a cubit wide, and three finger-breadths thick" with a grooved channel "a little more than one finger in breadth" cut along it—upon which he would roll, multiple times at multiple degrees of incline and lengths measured by a water clock and a pendulum, "a hard, smooth, and very round bronze ball" (*Discorsi*, 136–37).

16. Stengers, *The Invention of Modern Science*, 85.

17. Contrary to an enduring myth of modern science, experiments were common practice also before Galileo. It was their explanatory function that changed with Galileo's invention.

18. Galileo, *Discorsi*, 123–24.

19. Galileo, *Discorsi*, 190–96.

20. Michel Serres describes the political implications of this move: "Galileo is the first to put a fence around the terrain of nature, take it into his head to say, 'this belongs to science,' and find people simple enough to believe that this is of no consequence for man-made laws and civil societies. . . . The knowledge contract becomes identified with a new social contract. Nature then becomes global space, empty of men, from which society withdraws. . . . The experimental sciences make themselves masters of this empty, desert, savage space" (*Natural Contract*, 84–85).

21. Descartes, *Principles of Philosophy (Principia)*, IIP64. In subsequent

references from this text, I will quote in text the principles, following the same convention as for citing Spinoza's *Ethica*: here, IIP64 refers to part II, principle 64.

22. Already a well-established domain, statics in the seventeenth century actually comprised two quite different traditions—one based on the Aristotelian *Mechanica*, which meant that it formed part of Aristotle's natural philosophy, and another based on the purely mathematical work of Archimedes, notably pursued in the early seventeenth century by Simon Stevin. In the Aristotelian tradition, the scale measurement works in terms of a proportionality between weight and speed. In the Archimedean tradition, by contrast, bodies on a beam balance are treated as points along a line according to their center of gravity—that is, their physics is transformed into a mathematical model. Galileo's early work had dabbled in Aristotelian statics by trying to abstract physics from it but eventually abandoned it and rather pursued kinematics. Descartes drew more closely on Archimedean statics by trying to reintroduce physics into the mathematical model. For the general situation discussed in this chapter, however, this internal division of statics has little significance. Serres's *Birth of Physics* treats hydrostatics in the Archimedean vein and its relation to vortical motion in poetic detail. Gaukroger's account in *Emergence of a Scientific Culture* is more technically precise and leans more on the Aristotelian aspect.

23. Descartes, *The World*, 20–29.

24. Descartes, *The World*, 49.

25. As Gaukroger puts it, whereas Galileo's approach was to make physical questions amenable to mathematical treatment, "Descartes, by contrast, wants both to 'mathematize' physics and to 'physicalize' mathematics in one and the same operation. He does not simply want to use mathematics in physics, he wants to unify mathematics and physics in certain crucial respects" (Gaukroger, *Descartes: Philosophy*, 97–98).

26. Arendt, *The Human Condition*, 284.

27. Giorgio Agamben analyzes such a difference in some detail in *Language and Death*. He argues that this fracture in the field of being, between indication and signification, between showing and saying, "traverses the whole history of metaphysics, and without it, the ontological problem itself cannot be formulated. Every ontology (every metaphysics, but also every science that moves, whether consciously or not, in the field of metaphysics) presupposes the difference between indicating and saying, and is defined, precisely, as situated at the very limit between these two acts" (18). Agamben discusses the ontological difference between what he calls Voice and speech to its apotheosis: "As it

enacts the originary articulation of phone and logos through this double negativity, the dimension of the Voice constitutes the model according to which Western culture construes one of its own supreme problems: the relation and passage between nature and culture, between phusis and logos" (85).

28. For a concise philosophical argument on the direct implication of the Christian doctrine of creation for the possibility of a science on the modern model, see Michael B. Foster's articles from the 1930s, collected in Wybrow, *Creation, Nature, and Political Order.*

29. See note 21 regarding citations of the *Ethica.*

30. On the coherence of the first fifteen principles in Spinoza, see in particular Deleuze's article "Gueroult's General Method for Spinoza" (Gueroult's voluminous work has not been translated to English) in *Desert Islands,* 146–55.

31. The French philosopher Pierre Macherey is quite correct in observing that God in Spinoza's philosophy "is not 'one,' any more than he is two or three, or that he is beautiful or ugly. Contrary to a tenacious tradition, it must be said that Spinoza was no more a monist than he was a dualist, or a representative of any other number that one wants to assign to this fiction." Macherey quoted in Montag and Stolze, *The New Spinoza,* 88.

32. Stengers, *Cosmopolitics I,* 120–21.

33. Stengers, *Cosmopolitics I,* 106.

34. Stengers, *Cosmopolitics I,* 128.

35. Gaukroger, *Emergence of a Scientific Culture,* 505–7.

36. Although a similar "harmony" was crucial to previous natural philosophical orders, such as the Thomist synthesis under the Catholic Church, post-Reformation science profoundly rearranged this order according to a shifting faith. For instance, Gaukroger points to the Protestant inclination of the experimental natural philosophy that would become dominant with Newton, especially in its general emphasis on witnessing over received knowledge. "It is a core part of Protestant understanding . . . that unmediated access to the testimony of witnesses who were present at miraculous or otherwise holy events is to be preferred to the interpolations of generations of theologians" (*Emergence of a Scientific Culture,* 378).

3. Probability and Proliferation

1. Incidentally, there seems to be no clear etymological connection between *statics* and *statistics. Statics* comes from a Latin derivation of the Greek word for "weighing," whereas *statistics* is traced to a German invention of the nineteenth century, *Statistik,* likely connected to the word *state,* in the sense of state records.

2. Hacking, *Emergence of Probability*, 12.

3. Hacking, *Emergence of Probability*, 15.

4. Hacking, *Emergence of Probability*, 14.

5. Hacking, *Emergence of Probability*, 14.

6. Hacking, *Taming of Chance*, 2. Moreover, Agamben's *Homo Sacer* argues that biopolitics, defined by constitutive exception of a biological from a political dimension of life, strictly speaking is an ancient Western idea. In my general reference to the biopolitical in the following, I have in mind its modern, nineteenth-century appearance, since that is the purview of this chapter. However, this should not be read as a claim that biopolitics, or more specifically, metalogic, is necessarily a phenomenon exclusive to modern Western culture.

7. In this chapter I will largely be concerned with metalogic as statistical reason in the domain of physics, but that does not mean the metalogical is reducible to the effects of statistics. Consider briefly a remarkable historical parallel to the rise of probability and modern science: the joint-stock company, today known as the corporation. Also invented in mid-seventeenth-century Netherlands, corporations first became flourishing economic dynamos in western Europe in the same nineteenth-century period as concerns this chapter. As Joel Bakan puts it, "the genius of the corporation as a business form, and the reason for its remarkable rise over the last three centuries, was—and is—its capacity to combine the capital, and thus the economic power, of unlimited numbers of people" (*The Corporation*, 8). This combining capacity takes a very specific form. The decisive legal change that enabled the meteoric rise in corporate power throughout the Western world was first enacted in England in 1856: a statute of limited liability that simultaneously removed investor risk and constituted the corporation as a single juridical body. Thus: unlimited pooling of capital, combined in the legal entity of an individual, in which all causal connection between corporate owners and corporate actions has been formally removed. This new body is pure proliferation—the incorporation of leverage. Herein lies the essence of metalogic as I understand it.

8. Hacking, *Taming of Chance*, 2.

9. Hacking, *Taming of Chance*, 4.

10. In the academic literature on probability, this is a point that is curiously glossed over. In my view, this is because modern probability as a field of study and as a specialized branch of mathematics appears to take individual identity for granted. In Hacking's work too there is no problematization of how this identity must be forged if it doesn't already present itself as a given.

11. In the history of statistics, Hacking's cultural differentiation of East

and West has a corollary in the story of nineteenth-century physics. As Helge Kragh writes, it was mostly in Britain that "atomic models had their origin and were discussed. In Europe and in North America, interest in atomic structure was limited" (*Quantum Generations*, 48). In Germany toward the end of the nineteenth century, one of the most influential programs for unification of physics was "energeticism" and a rivaling conception of electromagnetic ontology founded on universal continuity. Similarly, many research programs in Britain led to J. J. Thomson's 1897 claim to having discovered the first particle, the electron, whereas in Berlin, Max Planck's mathematical result that light consists of discontinuous quanta—later canonized as the foundation of quantum theory—can at best be said to have been discovered accidentally, indirectly, and retrospectively.

12. Stengers, *Cosmopolitics I*, 206.

13. See Stengers's analysis of how Clausius effectively transforms the established physics of heat processes into a fictional device by use of hypothetical "state functions" and thereby establishing a system of equivalences through the principle of identity, in a similar vein to the discussion in the previous chapter. *Cosmopolitics I*, 206–8.

14. Maxwell, "Ether," 568–72.

15. P. M. Harman, *Energy, Force, and Matter*, 102.

16. Bergson, *Matter and Memory*, 266.

17. Bergson, *Matter and Memory*, 262–63.

18. Bergson, *Matter and Memory*, 265.

19. Bergson, *Creative Mind*, 169.

20. Bergson, *Matter and Memory*, 84.

21. Bergson, *Matter and Memory*, 3.

22. Bergson, *Matter and Memory*, 12.

23. Bergson, *Matter and Memory*, 14.

24. For Bergson, the relation between the extended and inextended was one of three related antitheses of perception, along with quality and quantity, freedom and necessity (*Matter and Memory*, 325). He avoided the term *intension*, as it was for him entangled with use in psychological discourse of the day. In Deleuze's reading of Bergson, and in the modulation of his own philosophy drawing on Bergsonian concepts, intension and intensity play prominent roles. My use of *intension* here is meant to generally illustrate the inverse tendency of extension, thought in terms of the concept of metalogic alone, which is not equivalent to Deleuze's usage of the term.

25. Bergson, *Matter and Memory*, 29–30.

26. Bergson, *Matter and Memory*, 32.

27. Bergson, *Matter and Memory*, 57.

28. Bergson, *Matter and Memory*, 15.

29. Bergson, *Matter and Memory*, 17.

30. Bergson, *Matter and Memory*, 68.

31. Bergson, *Matter and Memory*, 35.

32. Bergson, *Matter and Memory*, 196.

33. Bergson, *Matter and Memory*, 280.

34. Bergson, *Matter and Memory*, 239.

35. Bergson, *Matter and Memory*, 241–42.

36. Quoted in Krüger, "Slow Rise of Probabilism," 79.

37. In this sense, Maxwell's fictional local observer—he never himself used the term "demon"—must be distinguished from the more famous Laplace's demon, the omnidirectional observer from which the entire universe can be causally determined. Some histories even place Laplace's demon in the same category as Spinoza's alleged determinism. However, as I have shown in chapter 2, Spinoza's determinist perspective does not rely on an identifiable being. The idea of Laplace's demon, that a scientific observer could determine all movements of the universe, would be preposterous to Spinoza.

38. Kuhn, *Black-Body Theory*, 61.

39. Quoted in Lindley, *Boltzmann's Atom*, 199.

40. Kuhn, *Black-Body Theory*, 91.

41. Kuhn, *Black-Body Theory*, 70. Jan von Plato discusses the difference between classical and modern probability, defined as the departure from the equiprobability theorem of Huygens and Spinoza in favor of more autonomous mathematical approaches to calculation. See von Plato, "Probabilistic Physics the Classical Way," and *Creating Modern Probability*, esp. 4–18.

42. See Hacking, "Was There a Probabilistic Revolution," 52–53. Hacking's *Emergence of Probability* contains a discussion of "Cassirer's thesis" that the determinism of the 1870s is a distinctly different concept from the determinism retrospectively read back into seventeenth-century mechanics. In this sense, the supposed indeterminism of the late nineteenth century would be less of a historical break than is easily supposed when determinism and mechanism are conflated. In either case, my reading of Spinoza as the alleged archdeterminist of seventeenth-century thought, along with my discussion below, suggests the common historical distinction of determinism and indeterminism is a conceptual red herring, at least as long as they are considered on the same logical plane.

43. Nancy Cartwright has dedicated several books to this line of critique, including *How the Laws of Physics Lie*. Ilya Prigogine and Stengers's *The End of*

Certainty is a direct attempt at reestablishing the laws of physics on the statistical basis of thermodynamics. Prigogine's critical work is also reflected in Stengers's work, esp. *Power and Invention*, 21–79.

44. Campbell and Garnett, *Life of James Clerk Maxwell*, 439.

45. Kuhn, *Black-Body Theory*, 3. The following discussion is largely derived from Kuhn with some supplementary insight from Hentschel. I do not, however, follow Kuhn's historical conclusions on the significance of Planck's theory, nor am I concerned with his more controversial argument about the specific later sources of Planck's own reinterpretation of his own theory, something that fits Kuhn's general story of a paradigm shift but whose interpretation some historians of physics dispute.

46. Kuhn, *Black-Body Theory*, 112.

47. Kuhn, *Black-Body Theory*, 17–18.

48. Einstein, "On a Heuristic Point of View," 91–92.

49. In fact, all four of Einstein's "annus mirabilis" papers of 1905 contribute to establishing atomism in their own way. In his second paper, Einstein theorizes the movement of small particles suspended in a liquid, a phenomenon known from early-nineteenth-century microscopy as Brownian motion. Einstein shows how tiny pollen grains seen to be randomly moving around without any apparent cause can be mathematically explained as a generalized case of the kinetic theory of heat. From this, he reasons by analogy: if kinetic theory, which assumes atoms collide according to certain probabilistic rules, can be correctly used to calculate and predict the motion of visible particles, a correlation would hold for particles beyond the microscopic range, which are therefore subject to the same mathematical operation. Metaphysically, such an inference is by itself inadequate, since it does not preclude the physical possibility that atomic particles are mediated. In this respect, Einstein's third and fourth papers, which together establish the basis of special relativity, are significant, because they dispense with the prevailing conception of the autological ether through a simplification and integration of mechanist principles.

50. Quoted in Kragh, *Quantum Generations*, 41.

51. Falkenburg, *Particle Metaphysics*, 84–85.

52. On this point, see Milton, "Particle Physics," 455. In 2009 a research lab announced it had been able, through an advanced electron microscope, to show images of what it considered the structure of carbon atoms. Yet these are computer-regenerated reconstructions from so-called field emission electron microscopy, based on statistical modeling in which the atom is already implied. See Castelvecchi, "New Microscope."

53. On this point, see Egerton, *Physical Principles*, 20, and Bradbury, *Evolution of the Microscope*, 318.

54. Falkenburg, employing a very different terminology, nonetheless comes to a convergent conclusion with mine on the discrepancy, or mismatch, between dynamical and statistical modes of reasoning: "The core of the incommensurability problem in the transition from classical to quantum physics is the mismatch between the operational, axiomatic, and referential aspects of quantum concepts. This mismatch arises as follows. The data analysis of any high energy physics experiment forces physicists to analyze individual particle tracks and scattering events in quasi-classical terms. But from an axiomatic point of view, the operational basis of quantum mechanics and quantum field theory is probabilistic. Given that only classical point mechanics deals with individual particles, and given the unresolved quantum measurement problem at the level of individual particle detections, the mismatch is unavoidable" (*Particle Metaphysics*, 222).

55. Bergson, *Matter and Memory*, 259.

56. Falkenburg, *Particle Metaphysics*, 91.

57. Herein lies a direct historical and logical corollary between the constitution of physics and that of another metalogical enterprise—the corporation. As Bakan describes it: "By the end of the 19th century, through a bizarre legal alchemy, courts had fully transformed the corporation into a 'person,' with its own identity. . . . The corporate person had taken the place, at least in law, of the real people who owned corporations. Now viewed as an entity, 'not imaginary or fictitious, but real, not artificial but natural,' as it was described by one law professor in 1911, the corporation had been re-conceived as a free and independent being" (*The Corporation*, 16).

58. Canales, "Einstein, Bergson," 1169. Canales's interesting article recounts the scientific dispute between Einstein and Bergson in terms of a simultaneous political dispute between the two thinkers involving the League of Nations. The lingering bitterness between Einstein and Bergson after a series of confrontations, Canales suggests, left an indelible mark on Bergson's later years. In Bergson's final work, *The Two Sources of Morality and Religion*, his harmony of earlier years has decidedly given way to an ominous view of what he in this book called "the profound war instinct which covers civilization" (1182).

59. Canales, "Einstein, Bergson," 1171.

60. Deleuze, *Bergsonism*, 116.

4. Metaphysics with a Big Bang

1. Atkinson, "Herschel Space Telescope."

2. On the postwar history and figures, see Kragh, *Quantum Generations*, 295–310.

3. Stengers, *Cosmopolitics I*, 260.

4. Arendt, *Between Past and Future*, 9.

5. Arendt, *Between Past and Future*, 88.

6. The logic of a hypothesis that is made historically true in this sense must therefore not be confused with a tautology. For the hypothesis is not by itself a self-circular proposition. Rather, it is made circular, and thus foundational, through a forgetting of its conditions of emergence: the consistent action to which the axiomatic claim now owes its hegemonic existence.

7. Einstein, *Collected Papers*, 6:3. Note that when expressed this way, the principle of relativity is not actually relative at all. What it posits is rather an absolute and independent physical reality, irrespective of local conditions of understanding. The principle of relativity says, most essentially, "it is so," notwithstanding how it comes to be taken as so.

8. Einstein, *The Meaning of Relativity*, 30.

9. Einstein, *The Meaning of Relativity*, 30.

10. Einstein, *The Meaning of Relativity*, 28–29.

11. Einstein, *The Meaning of Relativity*, 56.

12. Einstein, *The Meaning of Relativity*, 58.

13. Einstein, *The Meaning of Relativity*, 58.

14. Lindley, *End of Physics*, 217.

15. Barrow, *Constants of Nature*, 24.

16. Barrow, *Constants of Nature*, 25.

17. In addition, there is the Coulomb constant, a proportionality governing electromagnetism used to formulate Planck charge. Its inverse square law makes it mathematically similar to gravitational G, and it retains its status as fundamental only insofar as electromagnetism and gravity are not theoretically unified—that is, its use is restricted. Finally, Boltzmann's infamous k, discussed in chapter 3, is derived from his statistical gas analysis and used nominally to formulate Planck temperature. However, in Planck's system of units, Boltzmann's k is a referential constant only, typically taking the value of 1.

18. Kragh, *Conceptions of Cosmos*, 138.

19. Einstein, *The Meaning of Relativity*, 107–8.

20. The following pages of overview of big bang cosmology up to the work of Stephen Hawking condenses a history that is told in myriad books.

For my brief account I have relied on Kragh, whose *Cosmology and Controversy* is the authoritative study of the battles between the steady state theory and the big bang theory. Throughout, these books have been complemented by insights from Narlikar, Lindley, and Seife.

21. From this premise, which alters the original four-dimensional space-time of the special theory of relativity, Friedmann was able to derive three possible models of the universe: positively curved, negatively curved, or, as the boundary condition between either of these two possibilities, flat. Temporally, these universes differ in their movement. Positive curvature implies a full closure—it expands, grows to a maximum size, then contracts. Both flat and negatively curved universes are open and expand forever, albeit at different rates. The current view in cosmology, based on observations, is that the likely structure of the universe is flat, meaning that it is continually expanding.

22. Kragh, *Conceptions of Cosmos*, 153.

23. Burbidge, "State of Cosmology," 4–5.

24. Kragh, *Conceptions of Cosmos*, 200–201.

25. In this sense, the light-particle, or the photon, is also a singularity, though considered a physical rather than mathematical construct.

26. Hawking, *Brief History of Time*, 52.

27. Hawking, *Brief History of Time*, 91.

28. Hawking, *Brief History of Time*, 52–53.

29. Disney, "Cosmology," 383.

30. Lindley, *End of Physics*, 11.

31. Guth, Hawking, et al., "A Cosmic Controversy."

32. Disney, "Cosmology," 384.

33. Disney, "Cosmology," 385.

34. Hacking, *Taming of Chance*, 56.

35. Planck, *Where Is Science Going?*, 28.

36. Pickering, *Constructing Quarks*, 214.

37. Hawking, *Brief History of Time*, 191.

38. Hawking, *Brief History of Time*, 128.

39. Carter, "Anthropic Principle in Cosmology."

40. Carter, "Anthropic Principle in Cosmology." The convention in physics discourse is now to distinguish between the "weak" and the "strong" versions of the anthropic principle. Their difference is reducible to the range of universal models in consideration. The weak version limits itself to saying that in an infinite or very large universe, conditions for intelligent life will only be met in very limited regions of such a universe. The strong version says that of all possible universes, only one like ours could produce such conditions. Whereas the

weak version is commonly accepted among physicists, the strong version has its detractors. See Carter, "Anthropic Principle in Cosmology"; and Hawking, *Brief History of Time*, 128–29.

41. Hawking, *Brief History of Time*, 128–29.

Conclusion

1. Stengers, *Cosmopolitics*, 9.
2. Stengers, *Cosmopolitics*, 47.
3. Latour, *Pandora's Hope*, 305.
4. Stengers, *Power and Invention*, 13.
5. Stengers, *Power and Invention*, 6.
6. Stengers, *Power and Invention*, 6.

BIBLIOGRAPHY

Agamben, Giorgio. *Homo Sacer: Sovereign Power and Bare Life*. Trans. Daniel Heller-Roazen. Stanford: Stanford University Press, 1998.

———. *Language and Death: The Place of Negativity*. Trans. Karen E. Pinkus and Michael Hardt. Minneapolis: University of Minnesota Press, 1991.

Aiton, E. J. *The Vortex Theory of Planetary Motions*. London: Macdonald, 1972.

Albir, Toni. "Physicists Increasingly Confident They've Found the Higgs Boson." *National Geographic*, March 15, 2013, online edition.

Arendt, Hannah. *Between Past and Future: Eight Exercises in Political Thought*. New York: Penguin Books, 2006.

———. *The Human Condition*. 2nd ed. Chicago: University of Chicago Press, 1998.

Atkinson, Nancy. "Herschel Space Telescope Closes Its Eyes on the Universe." *Universe Today*, April 29, 2013.

Badiou, Alain. *Deleuze: The Clamor of Being*. Minneapolis: University of Minnesota Press, 2000.

Baggott, Jim. *Farewell to Reality: How Modern Physics Has Betrayed the Search for Scientific Truth*. Cambridge, UK: Pegasus, 2013.

Bakan, Joel. *The Corporation: The Pathological Pursuit of Profit and Power*. Toronto: Viking Canada, 2004.

Bal, Hartosh Singh. "Fundamental Forces and Chopping Wood." *Open Magazine*, February 13, 2010.

Barrow, John D. *The Constants of Nature: From Alpha to Omega*. London: Jonathan Cape, 2002.

Beistegui, Miguel de. *Truth and Genesis: Philosophy as Differential Ontology.* Bloomington: Indiana University Press, 2004.

Bergson, Henri. *Creative Evolution.* Trans. Arthur Mitchell. New York: Holt, 1911.

———. *The Creative Mind.* New York: Philosophical Library, 1946.

———. *Duration and Simultaneity: With Reference to Einstein's Theory.* Indianapolis: Bobbs-Merrill, 1965.

———. *Matter and Memory.* London: G. Allen & Unwin, 1911.

Biagioli, Mario. *Galileo, Courtier: The Practice of Science in the Culture of Absolutism.* Chicago: University of Chicago Press, 1993.

Billings, Lee. "The Telescope That Ate Astronomy." *Nature,* October 28, 2010, 1028–30.

Boslough, John. *Stephen Hawking's Universe: An Introduction to the Most Remarkable Scientist of Our Time.* New York: Avon, 1989.

Bradbury, Savile. *The Evolution of the Microscope.* Oxford: Pergamon Press, 1967.

Burbidge, Geoffrey. "The State of Cosmology." In *Current Issues in Cosmology,* ed. Jean-Claude Pecker and Jayant Narlikar, 3–16. Cambridge: Cambridge University Press, 2006.

Campbell, Lewis, and James Garnett. *The Life of James Clerk Maxwell: With a Selection from His Correspondence.* Cambridge: Cambridge University Press, 2010.

Canales, Jimena. "Einstein, Bergson, and the Experiment That Failed: Intellectual Cooperation at the League of Nations." *Modern Language Notes,* no. 120 (2005): 1168–91.

Capek, Milic. *Bergson and Modern Physics: A Reinterpretation and Re-Evaluation.* Dordrecht: Reidel, 1971.

Capra, Fritjof. *The Tao of Physics: An Exploration of the Parallels between Modern Physics and Eastern Mysticism.* 2nd ed. Boston: New Science Library, 1983.

Carroll, Sean. *The Particle at the End of the Universe.* London: Oneworld Books, 2012.

Carter, Brandon. "Anthropic Principle in Cosmology." 2006. Contribution to the colloquium "Cosmology: Facts and Problems." College de France, June 2004. Published online, https://arxiv.org/abs/gr-qc/0606117v1.

Cartwright, Nancy. *How the Laws of Physics Lie.* Oxford: Oxford University Press, 1983.

Cassirer, Ernst. *The Problem of Knowledge; Philosophy, Science, and History since Hegel.* New Haven: Yale University Press, 1960.

Castelvecchi, Davide. "New Microscope Reveals the Shape of Atoms." *Scientific American,* December 1, 2009.

Chalmers, Matthew. "Stringscape." *Physics World*, September 2007, 35–47.

Cohen, I. Bernard. "Scientific Revolutions, Revolutions in Science, and a Probabilistic Revolution, 1800–1930." In *The Probabilistic Revolution*, vol. 1, *Ideas in History*, ed. Lorenz Krüger, Lorraine J. Daston, and Michael Heidelberger, 23–44. Cambridge: MIT Press, 1987.

Deleuze, Gilles. *Bergsonism*. Trans. Hugh Tomlinson and Barbara Habberjam. New York: Zone Books, 1988.

———. *Desert Islands and Other Texts, 1953–1974*. Trans. Michael Taormina. Los Angeles: Semiotext(e), 2004.

———. *Spinoza, Practical Philosophy*. Trans. Robert Hurley. San Francisco: City Lights Books, 1988.

Descartes, René. *The Philosophical Writings of Descartes*. Vol. 1. Trans. John Cottingham, Robert Stoothof, and Dugald Murdoch. Cambridge: Cambridge University Press, 1985.

———. *Principles of Philosophy*. Trans. R. P. Miller. Dordrecht, The Netherlands: Springer NL/Reidel, 1982.

———. *The World and Other Writings*. Trans. and ed. Stephen Gaukroger. Cambridge: Cambridge University Press, 1998.

Disney, Michael J. "The Case against Cosmology." *General Relativity and Gravitation*, no. 32 (2000): 1125–34.

———. "Cosmology: Science or Folk Tale?" *American Scientist* 95, no. 5 (September–October 2007): 383–85.

Egerton, R. F. *Physical Principles of Electron Microscopy: An Introduction to Tem, Sem, and Aem*. New York: Springer, 2005.

Einstein, Albert. *The Collected Papers of Albert Einstein*. 14 vols. Princeton, N.J.: Princeton University Press, 1987.

———. *The Meaning of Relativity: Including the Relativistic Theory of the Non-Symmetric Field*. Princeton, N.J.: Princeton University Press, 2005.

———. "On a Heuristic Point of View about the Creation and Conversion of Light." *Annalen der Physik* 17, no. 132 (1905): 91–107.

Evans, Robert. "Higgs 'God' Particle a Big Let-Down Say Physicists." *Globe and Mail*, March 8, 2013, online edition.

Falkenburg, Brigitte. *Particle Metaphysics: A Critical Account of Subatomic Reality*. Heidelberg: Springer, 2007.

———. "Scattering Experiments." In *Compendium of Quantum Physics: Concepts, Experiments, History, and Philosophy*, ed. Daniel Greenberger, Klaus Hentschel, and Friedel Weinert, 676–81. Berlin: Springer, 2009.

Folse, Henry J. *The Philosophy of Niels Bohr: The Framework of Complementarity*. New York: Elsevier, 1985.

Ford, Russell. "Immanence and Method: Bergson's Early Reading of Spinoza." *Southern Journal of Philosophy* 42 (2004): 171–92.

Foucault, Michel. *The Order of Things: An Archaeology of the Human Sciences.* London: Tavistock, 1970.

Friedman, Michael. *A Parting of the Ways: Carnap, Cassirer, and Heidegger.* Chicago: Open Court, 2000.

Galilei, Galileo. *Dialogue concerning the Two Chief World Systems: Ptolomaic & Copernican.* Trans. Stillman Drake. Rev. ed. Berkeley: University of California Press, 1967.

———. *Dialogues concerning Two New Sciences/by Galileo Galilei.* Trans. Henry Crew. Philadelphia: Running Press, 2002.

Gaukroger, Stephen. *Descartes: An Intellectual Biography.* Oxford: Oxford University Press, 1995.

———. *Descartes: Philosophy, Mathematics and Physics.* Brighton, Sussex: Harvester Press, 1980.

———. *Descartes' System of Natural Philosophy.* Cambridge: Cambridge University Press, 2002.

———. *The Emergence of a Scientific Culture: Science and the Shaping of Modernity, 1210–1685.* Oxford: Clarendon Press, 2006.

———. *Explanatory Structures: A Study of Concepts of Explanation in Early Physics and Philosophy.* Hassocks: Harvester Press, 1978.

Gellner, Ernest. *Society and Western Anthropology.* New York: Columbia University Press, 1980.

Guth, Alan, Stephen Hawking, et al. "A Cosmic Controversy: A Reply to the Critics." *Scientific American,* March 2017.

Gutting, Gary. *Continental Philosophy of Science.* Malden, Mass.: Blackwell, 2005.

Hacking, Ian. *The Emergence of Probability: A Philosophical Study of Early Ideas about Probability, Induction, and Statistical Inference.* 2nd ed. New York: Cambridge University Press, 2006.

———. *Representing and Intervening: Introductory Topics in the Philosophy of Natural Science.* Cambridge: Cambridge University Press, 1983.

———. *The Taming of Chance.* Cambridge: Cambridge University Press, 1990.

———. "Was There a Probabilistic Revolution 1800–1930?" In *The Probabilistic Revolution,* vol. 1, *Ideas in History,* ed. Lorenz Krüger, Lorraine J. Daston, and Michael Heidelberger, 45–58. Cambridge: MIT Press, 1987.

Harman, Graham. *Prince of Networks: Bruno Latour and Metaphysics.* Melbourne: Re.Press, 2009.

Harman, P. M. *Energy, Force, and Matter: The Conceptual Development of Nineteenth-Century Physics*. Cambridge: Cambridge University Press, 1982.

———. *Metaphysics and Natural Philosophy: The Problem of Substance in Classical Physics*. Totowa, N.J.: Harvester Press, 1982.

Hart, Matthew. "A Gleam in God's Eye." *Globe and Mail*, December 23, 2006, F1, 6–7.

Hawking, S. W. *A Brief History of Time*. Updated and expanded tenth anniversary ed. New York: Bantam Books, 1998.

———. *Is the End in Sight for Theoretical Physics?* Cambridge: Cambridge University Press, 1980.

Heidegger, Martin. *The Basic Problems of Phenomenology*. Bloomington: Indiana University Press, 1988.

———. *Basic Writings: From "Being and Time" (1927) to "The Task of Thinking" (1964)*. Ed. David Farrell Krell. Rev. and expanded ed. San Francisco: HarperSanFrancisco, 1993.

———. *Identity and Difference*. Trans. Joan Stambaugh. Chicago: University of Chicago Press, 2002.

———. *The Principle of Reason*. Bloomington: Indiana University Press, 1996.

———. *The Question concerning Technology, and Other Essays*. New York: Harper & Row, 1977.

Heisenberg, Werner. *Physics and Philosophy: The Revolution in Modern Science*. New York: Harper, 1958.

Hentschel, Klaus. "Light Quantum." In *Compendium of Quantum Physics: Concepts, Experiments, History, and Philosophy*, ed. Daniel Greenberger, Klaus Hentschel, and Friedel Weinert, 339–46. Berlin: Springer, 2009.

Holton, Gerald James. *Einstein and His Perception of Order in the Universe*. Ottawa: Carleton University, 1979.

Husserl, Edmund. "The Crisis of European Sciences and Transcendental Phenomenology." In *Continental Philosophy of Science*, ed. Gary Gutting, 113–20. Malden, Mass.: Blackwell, 2005.

Ijjas, Anna, Paul J. Steinhardt, and Abraham Loeb. "Pop Goes the Universe: Cosmic Inflation Theory Faces Challenges." *Scientific American*, February 2017.

Jameson, Fredric. *A Singular Modernity: Essay on the Ontology of the Present*. London: Verso, 2002.

Jammer, Max. *Concepts of Force: A Study in the Foundations of Dynamics*. Cambridge: Harvard University Press, 1957.

Kolbert, Elizabeth. "Crash Course." *New Yorker*, May 14, 2007.

Kragh, Helge S. *Conceptions of Cosmos: From Myths to the Accelerating Universe: A History of Cosmology*. Oxford: Oxford University Press, 2007.

——. *Cosmology and Controversy: The Historical Development of Two Theories of the Universe*. Princeton, N.J.: Princeton University Press, 1996.

——. *Quantum Generations: A History of Physics in the Twentieth Century*. Princeton, N.J.: Princeton University Press, 1999.

Krauss, Lawrence M. *A Universe from Nothing: Why There Is Something Rather Than Nothing*. New York: Atria Books, 2013.

Krüger, Lorenz. "The Probabilistic Revolution in Physics—An Overview." In *The Probabilistic Revolution*, vol. 2, *Ideas in the Sciences*, ed. Lorenz Krüger, Gerd Gigerenzer, and Mary S. Morgan, 373–78. Cambridge: MIT Press, 1990.

——. "The Slow Rise of Probabilism: Philosophical Arguments in the Nineteenth Century." In *The Probabilistic Revolution*, vol. 1, *Ideas in History*, ed. Lorenz Krüger, Lorraine J. Daston, and Michael Heidelberger, 59–90. Cambridge: MIT Press, 1987.

Kuhn, Thomas S. *Black-Body Theory and the Quantum Discontinuity, 1894–1912*. Oxford: Oxford University Press, 1978.

——. *The Structure of Scientific Revolutions*. 3rd ed. Chicago: University of Chicago Press, 1996.

——. "What Are Scientific Revolutions?" In *The Probabilistic Revolution*, vol. 1, *Ideas in History*, ed. Lorenz Krüger, Lorraine J. Daston, and Michael Heidelberger, 7–22. Cambridge: MIT Press, 1987.

Latour, Bruno. *Modes of Existence*. Cambridge: Harvard University Press, 2013.

——. *Pandora's Hope: Essays on the Reality of Science Studies*. Cambridge: Harvard University Press, 1999.

——. *Politics of Nature: How to Bring the Sciences into Democracy*. Cambridge: Harvard University Press, 2004.

——. "'Thou Shall Not Freeze-Frame'; or How Not to Misunderstand the Science and Religion Debate." In *Science, Religion, and the Human Experience*, ed. James D. Proctor, 27–48. Oxford: Oxford University Press, 2005.

——. *We Have Never Been Modern*. Cambridge: Harvard University Press, 1993.

Lederman, Leon M., and Dick Teresi. *The God Particle: If the Universe Is the Answer, What Is the Question?* Boston: Houghton Mifflin, 1993.

Leibniz, Gottfried Wilhelm. *Philosophical Texts*. Trans. and ed. R. S. Woolhouse and Richard Francks. Oxford University Press, 1998.

Lindley, David. *Boltzmann's Atom: The Great Debate That Launched a Revolution in Physics*. New York: Free Press, 2001.

————. *The End of Physics: The Myth of a Unified Theory*. New York: Basic-Books, 1993.

————. *Uncertainty: Einstein, Heisenberg, Bohr, and the Struggle for the Soul of Science*. New York: Doubleday, 2007.

Mainzer, K. "Symmetry." In *Compendium of Quantum Physics: Concepts, Experiments, History, and Philosophy*, ed. Daniel Greenberger, Klaus Hentschel, and Friedel Weinert, 779–85. Berlin: Springer, 2009.

Maxwell, James Clerk. "Ether." *Encyclopedia Britannica*, 9th ed., 568–72. London, 1878.

Milton, Kimball. "Particle Physics." In *Compendium of Quantum Physics: Concepts, Experiments, History, and Philosophy*, ed. Daniel Greenberger, Klaus Hentschel, and Friedel Weinert, 455–59. Berlin: Springer, 2009.

Montag, Warren, and Ted Stolze. *The New Spinoza*. Minneapolis: University of Minnesota Press, 1997.

Montebello, Pierre. "Matter and Light in Bergson's *Creative Evolution*." *SubStance* 36, no. 3 (2007).

Narlikar, Jayant Vishnu. *Introduction to Cosmology*. 3rd ed. Cambridge: Cambridge University Press, 2002.

Newton, Isaac, Andrew Motte, and N. W. Chittenden. *Newton's Principia: The Mathematical Principles of Natural Philosophy*. 1st American ed. New York: D. Adee, 1848.

Nietzsche, Friedrich Wilhelm, and Walter Arnold Kaufmann. *The Gay Science; with a Prelude in Rhymes and an Appendix of Songs*. New York: Random House, 1974.

Parsons, Keith, ed. *The Science Wars: Debating Scientific Knowledge and Technology*. New York: Prometheus Books, 2003.

Pecker, Jean Claude, and Jayant Vishnu Narlikar. *Current Issues in Cosmology*. Cambridge: Cambridge University Press, 2006.

Pickering, Andrew. *Constructing Quarks: A Sociological History of Particle Physics*. Chicago: University of Chicago Press, 1984.

Planck, Max. *Where Is Science Going?* Trans. and ed. James Murphy. New York: Norton, 1932.

Polkinghorne, J. C. *Quantum Physics and Theology: An Unexpected Kinship*. New Haven: Yale University Press, 2007.

Prigogine, Ilya, and Isabelle Stengers. *The End of Certainty: Time, Chaos, and the New Laws of Nature*. New York: Free Press, 1997.

Rohrlich, Daniel. "Errors and Paradoxes in Quantum Mechanics." In *Compendium of Quantum Physics: Concepts, Experiments, History, and Philosophy*, ed. Daniel Greenberger, Klaus Hentschel, and Friedel Weinert, 211–20. Berlin: Springer, 2009.

Saunders, Simon. "Identity of Quanta." In *Compendium of Quantum Physics: Concepts, Experiments, History, and Philosophy*, ed. Daniel Greenberger, Klaus Hentschel, and Friedel Weinert, 299–304. Berlin: Springer, 2009.

Seife, Charles. *Alpha & Omega: The Search for the Beginning and End of the Universe*. New York: Viking, 2003.

Serres, Michel. *The Birth of Physics*. Trans. Jack Hawkes. Manchester: Clinamen, 2000.

———. *The Natural Contract: Studies in Literature and Science*. Ann Arbor: University of Michigan Press, 1995.

———. *The Parasite*. Minneapolis: University of Minnesota Press, 2007.

———. "Revisiting the Natural Contract." *CTheory, Thousand Days of Theory: 039*, November 15, 2006.

Serres, Michel, and Bruno Latour. *Conversations on Science, Culture, and Time*. Trans. Roxanne Lapidus. Ann Arbor: University of Michigan Press, 1995.

Sloterdijk, Peter. *Eurotaoismus: Zur Kritik der politischen Kinetik*. Frankfurt am Main: Suhrkamp, 1989.

———. *Terror from the Air*. Trans. Amy Patton and Steve Corcoran. Cambridge, Mass.: Semiotext(e), 2009.

Smolin, Lee. *The Trouble with Physics: The Rise of String Theory, the Fall of a Science, and What Comes Next*. New York: Mariner Books, 2007.

Spinoza, Benedictus de. *Ethics*. Trans. Edwin Curley. London: Penguin Books, 1996.

———. *Spinoza's Algebraic Calculation of the Rainbow; &, Calculation of Chances*. Trans. and ed. Michael John Petry. Dordrecht: M. Nijhoff, 1985.

Stenger, Victor J. "Quantum Metaphysics." *Scientific Review of Alternative Medicine* 1, no. 1 (1997): 26–30.

Stengers, Isabelle. *Cosmopolitics I*. Trans. Robert Bononno. Minneapolis: University of Minnesota Press, 2010.

———. *The Invention of Modern Science*. Trans. Daniel W. Smith. Minneapolis: University of Minnesota Press, 2000.

———. *Power and Invention: Situating Science*. Trans. Paul Bains. Minneapolis: University of Minnesota Press, 1997.

Stewart, Matthew. *The Courtier and the Heretic: Leibniz, Spinoza, and the Fate of God in the Modern World*. New York: Norton, 2006.

Trusted, Jennifer. *Physics and Metaphysics: Theories of Space and Time*. London: Routledge, 1991.

Veneziano, Gabriele. "The Myth of the Beginning of Time." *Scientific American*, May 2004, 54–65.

von Plato, Jan. *Creating Modern Probability: Its Mathematics, Physics, and Philosophy in Historical Perspective.* Cambridge: Cambridge University Press, 1994.

———. "Probabilistic Physics the Classical Way." In *The Probabilistic Revolution*, vol. 2, *Ideas in the Sciences*, ed. Lorenz Krüge, Gerd Gigerenzer, and Mary S. Morgan, 379–408. Cambridge: MIT Press, 1987.

Wheaton, Bruce R. "Wave-Particle Duality; Some History." In *Compendium of Quantum Physics: Concepts, Experiments, History, and Philosophy*, ed. Daniel Greenberger, Klaus Hentschel, and Friedel Weinert, 830–40. Berlin: Springer, 2009.

Whitehead, Alfred North. *The Concept of Nature. The Tarner Lectures: 1919.* Cambridge: Cambridge University Press, 1920.

———. *Process and Reality: An Essay in Cosmology.* New York: Free Press, 1979.

———. *Science and the Modern World.* Cambridge: Cambridge University Press, 1933.

Witten, Edward. "Universe on a String." *Astronomy*, June 2002, 41–47.

Woit, Peter. *Not Even Wrong: The Failure of String Theory and the Search for Unity in Physical Law.* New York: Basic Books, 2007.

Wolchover, Natalie. "A Fight for the Soul of Science." *Quanta*, December 16, 2015.

Wolfe, Cary. *What Is Posthumanism?* Minneapolis: University of Minnesota Press, 2010.

Wybrow, Cameron, ed. *Creation, Nature, and Political Order in the Philosophy of Michael Foster (1903–1959): The Classic Mind Articles and Others, with Modern Critical Essays.* Lewiston, N.Y.: E. Mellen Press, 1992.

INDEX

fine structure, 152; gravitation G, 90; invention of, 56, 59, 62, 76; Planck, 35, 133–37, 151–53, 165–66

constructivism, 9, 24–27. *See also* realism

cosmology: as branch of physics, 26–28, 32–35, 38, 134–38; Cartesian, 57–61; current scientific, 1–6, 124–26, 147–52, 160–66; emergence of current, 139–47; history of, 10–13, 23, 45, 82, 90, 116; as metaphysics, 18–21, 153–57. *See also* big bang theory

cosmos, as distinct from universe, 11–12, 51, 120, 137, 155–60

dark matter, 5, 150–51, 157

Deleuze, Gilles, 9, 103, 117, 173n30, 175n24

Descartes, René, 13, 44–48, 57–76; vs. Bergson, 96–97, 102; on cogito, 72–75; historical role, 47–48, 76; on *mathesis universalis*, 34; vs. Spinoza, 67–73; vortex cosmology, 57–62, 93, 156

determinism, vs. indeterminism, 100–101, 176n37, 176n42

Disney, Michael J., 5, 147–51

dynamics: classical, 44–45, 50–53, 57, 60–62, 67, 75; confused with metalogic, 113, 115, 137, 178n54; vs. thermodynamics, 90–92, 105–8

E=mc², 127–28, 135–36. *See also* general relativity theory

Einstein, Albert, 89, 111–12, 122–23; debate with Bergson, 117–18;

link to Hawking, 145–48. *See also* general relativity theory

electron, 104, 111–16, 136

electron microscope, 114–15, 177n52

empiricism, 18, 24–27, 149

energy, 33, 50, 90–93, 104–5, 108–11; relation to matter, 114–15, 127–29, 132, 135–36, 144

entropy, 90–91, 104–5, 107

equilibrium, 49, 58, 64, 74–75; in steady state theory, 142; in thermodynamics, 91, 106–10. *See also* statics

equivocal, 66–69, 73, 77. *See also* univocal

ether, 36, 93–95, 156, 177n49

event, 24–31, 128, 141–43, 147

experiment: role of, 9–10, 21, 28–40, 53, 113, 164–66; of thought, 63, 105, 131

faith, in physics, 12, 18–20, 125, 151–53, 153–57, 164

Falkenburg, Brigitte, 25–27, 113, 116

forces: in Descartes and Newton, 55–60: in Galileo, 50–52; in Lagrange, 74–75; problem of, 32–36, 130–33; in thermodynamics, 90–95. *See also* gravity

free space. *See* vacuum

free will, 63, 71, 100, 107

Friedmann, Alexander, 140, 145–46, 180n21

fundamental laws. *See* theoretical laws

Galileo, Galilei, 13, 42–46, 50, 61–76, 152, 157; as forerunner of Newton, 57, 65, 71–76, 160; as

Mach, Ernst, 19–20
mathematical, idea of the, 34–36,
  39–40, 50–52, 65
mathematical unification. See *mathesis universalis*
mathematization: of metaphysics,
  35–36, 145–47, 151–57; of physics,
  11, 50–57, 74–77, 92, 109–13
mathesis universalis, 34, 38–40, 61,
  73–74, 96, 153–57
matter, 32, 58–61, 65, 91–92, 94–96;
  equivalence with energy, 127–29;
  as particulate, 112–16; in the universe, 139, 142, 144–47. *See also*
  dark matter; energy
Maxwell, James Clerk, 13, 89–95,
  105–8, 112, 161
mechanics: classical, 51, 75–76,
  92–93; Einstein's critique of,
  130–31; reversibility of, 107–8,
  146
mediation, 31, 47, 72, 170n19; autological, 93–95, 102, 165; removal
  of, 56, 63
metalogic, 14, 86–89, 161–63, 174n7,
  175n57; as revolution in physics,
  104–12, 114–19, 125, 134–35
metaphysics, 1–2, 8–14, 17–18,
  38–40, 153–57; actual use in physics, 26–27, 31–32, 37, 51–52; battle
  of, 124–26, 141–45; vs. Bergsonist,
  101; Cartesian, 61–67, 76; of
  general relativity, 127–33, 145–46;
  principles of, 48–50, 73–74;
  reconciliation with science, 95,
  103–4, 107–8, 117–19; revolution
  of, 77–78, 82–83, 105, 111, 115; vs.
  Spinozist, 68; in thermodynamics, 90–92, 105; as transcendental

idea, 20–21, 164; as "unscientific,"
  18, 25, 37, 149–51
microscope: electron type, 114–15,
  177n52; historical role of, 23–24,
  27, 35–36, 40, 60, 81–83
microwave background radiation,
  138, 143

natural philosophy, 45, 50–52, 65–67,
  76–77
natural theology, 50, 76, 108
Neptune, 80–83
Newton, Isaac, 13, 23, 25, 39, 57–60,
  160; dynamics (mechanics), 45,
  51–52, 107–8, 130; on gravity
  (G), 8, 32, 59, 81, 132–35; idea of
  universe, 74–77, 89–92, 127–28,
  136–39; link to Galileo, 54–55,
  65, 67
nucleosynthesis. *See* atom; big bang
  theory

ontology: classical, 77, 172n27;
  vs. metaphysics, 12, 27, 116;
  positivist, 85. *See also* metaphysics

paradigm, current cosmological,
  3–6, 13, 40, 147, 160–64
particle, 25–27, 32, 35–38, 94–95,
  109; as basis for big bang theory, 140–44; invention of, 82–83,
  112–16
phenomenological laws (physics), 7,
  34, 107, 152
phenomenology (philosophy), 118
Planck, Max, 19–20, 106, 109–12,
  128–29, 153
Planck constant, 35, 133–37, 151–53,
  165–66

68, 72, 164; as telos in science, 10, 19–20, 29, 53, 156–57, 165–66. *See also* belief

universe: Bergsonian, 96–98, 102; composition of, 32, 35–38, 147–52; as distinct from God, 65–67; Einsteinian, 129–32, 137–41; as expanding, 139–41, 145–48; idea of, 1–2, 10–13, 51, 77, 82, 160–65; mathematical treatment of, 34, 39–40, 50–52, 74, 131–33, 151–52; as mirror of humanism, 88, 153–56; Newtonian, 51–52, 55, 57, 127; as particle accelerator, 21, 38, 144; scale of, 137–38; in thermodynamics, 90–92. *See also* cosmology: Cartesian; general relativity theory

universe, new scientific model of. *See* big bang theory

univocal, 65–66, 69, 73, 97, 101. *See also* equivocal

vacuum, 57, 112, 128–31. *See also* speed of light

void, 49–50, 54–57, 65–67, 90–91, 95–96, 160

vortex theory, 60–61, 67, 76, 93–94, 156

world-alienation, 44, 159–60

world-object, 1–2, 14, 17, 21, 44, 166

world-picture (current cosmological), 17–21, 35, 155–57, 162–64; Cartesian, 57–65; modern (Bergson), 104; Newtonian, 51; Spinozist, 68–70

(continued from page ii)

Bjørn Ekeberg has a PhD in cultural, social, and political thought from the University of Victoria, Canada, and is an independent researcher on philosophy, science, media, and politics. He lives in Oslo, Norway.